Protecting Paradise

300 Ways to Protect Florida's Environment

by

PEGGY CAVANAUGH

and

MARGARET SPONTAK

published by
PHOENIX PUBLISHING

This book is dedicated to Florida's children.

• • •

• • •

• • •

Cover design, layout and inside artwork by
Michael L. Williams Jr.

• • •

Photography by Peggy Cavanaugh

• • •

• • •

For ordering information, write to:

Phoenix Publishing
P.O. Box 866
Fairfield, Florida 32634

Bulk rates are available.

• • •

Text Printed on 100 percent recycled stock and printed with soybean inks.

Printed in the United States of America.

FOREWORD

Here is a book that will bring joy and comfort to many worried Floridians. It tells in a simple straightforward way how they can do something to protect the land they live in.

There is an almost limitless number of books published describing how to cook every sort of viand, how to sew and knit and embroider and quilt. There are books on how to build swimming pools, boats, log cabins and dog houses. You can find printed directions on how to dry fruits and vegetables, grow roses, install wall paper and play cribbage. But this is the first book I know of that describes for the ordinary, plain everyday citizen how to protect his environment—his Paradise.

The authors are positive in their overlook. They present a brief sprightly description of each of the many threats of degradation to the environment and immediately suggests some possible steps to be taken. Success stories dealing with similar situations encourage the reader to take action. Finally there are additional references dealing with the particular issue and sources for more information and assistance. Altogether, it is a most valuable book for the growing army of Florida citizens who are becoming serious and effective stewards for the Florida environment.

It is now abundantly evident that if our Paradise is to be protected, the initiative and force will have to come from the grassroots. The action will have to start at the local level and move upward. That is natural and proper in a democracy. Many of the institutions we value and take for granted, such as schools, started at the local village level in the early days and then spread to the state and national level . So it must be with the newly defined environmental ethic.

Here in Florida we have indicated over and over again that the public places great value on the preservation of original landscapes, wildlife and ground water. The public wants good stewardship practiced for all facets of the environment, and more to the point, it is willing to pay for it.

So why aren't we more successful when we assess our local conditions at the beginning of the 1990's? I think it is because we have looked for solutions to these local problems from leadership that is far away. It doesn't work. It is the citizen who must lead; and to lead effectively the citizen needs to have certain basic knowledge. This book, *Protecting Paradise* provides that basic information . Read it, enjoy it, keep it close by for easy reference, and you will be a happier and less frustrated citizen of Florida.

> Marjorie Harris Carr
> Gainesville, Florida
> February 1992

Special note: Marjorie H. Carr is one of the founders of Florida Defenders of the Environment (FDE) and served as president for sixteen years. The group began as a coalition of volunteer specialists, including scientists, lawyers and other professionals, who helped halt the Cross Florida Barge Canal.

Today, FDE still utilizes volunteer experts to provide reliable information on Florida environmental issues to decision makers. Using sound research has made FDE an organization that is recognized and respected by elected officials, educators, environmentalists, government agencies, media, and corporate executives.

Some of FDE's current projects include protecting rivers such as the Apalachicola, Oklawaha and Suwanee; researching toxic issues; establishing local environmental networks; and participating in the Everglades Coalition.

A portion of the profits from this book will be donated to FDE.

Special thanks to the following people who helped provide direction and input in developing **Protecting Paradise.**

• • •

Mark Benedict, Florida Greenways Project
Marjorie Carr, Florida Defenders of the Environment
Kerry Dressler, Florida Defenders of the Environment
John Hankinson, St. Johns River Water Management District
Shirley Hankinson, Retired Teacher and School Administrator
Ruth Heaton, St. Johns River Water Management District
Clay Henderson, Volusia County Land Acquisition
Reid Hughes, Florida Environmental Education Foundation
Connie Jones, Independent Information Services, Inc.
Charles Lee, Florida Audubon
Shirley Little, Florida Defenders of the Environment
Buddy MacKay, Florida Lt. Governor
Patti McKay, 1000 Friends of Florida
Nathaniel P. Reed, 1000 Friends of Florida
Bob Simons, Florida Defenders of the Environment
Kent Wimmer, 1000 Friends of Florida
and all the environmental organizations and
individuals that sent us materials on request.

• • •

We also want to thank Gene McConnell and
Michael L. Williams Jr. of Special Publications, Inc.
for their dedication to excellence in design and production.

• • •

Of course, we thank all our friends and family for their
patience throughout this endless process.

*Bravo on **Protecting Paradise**!
It's needed, it's worthy of a "sellout"
and it needs to be required reading
in Florida's school system.*

*Nathaniel P. Reed
Founder and President
1000 Friends of Florida*

CONTENTS

FOREWORD BY MARJORIE CARR

THINK GLOBALLY, ACT LOCALLY

"The best and often the only way to deal with global problems is, paradoxically, to look for solutions peculiar to each locality...If we really want to do something for our planet, the best place to start is in the streets, fields, roads, rivers, marshes, and coastlines of our communities."

—René Dubos

For Floridians that advice takes on special meaning. Ecosystems in our state have global significance. The Everglades has been recognized by the United Nations as an International Biosphere Reserve. The Florida Keys, the only tropical ecosystem in the continental United States has been designated as one of 12 "Last Great Places" in the world by the Nature Conservancy. The Archie Carr National Wildlife Refuge on the east coast of Florida is the world's second largest loggerhead turtle rookery.

We could talk also of Silver Springs, with its flow greater than all but one other freshwater spring in the world, or of our ancient scrub, or mangrove swamps which serve as nurseries for so many forms of marine life.

Florida is number one in the nation for diversity of trees and lakes and a close second in miles of coastline. Florida has more than 26 million acres of state and privately owned forest and wildlands and some 38,000 species of plants. We are home to some of the most unique endangered species from the Florida panther to the manatee.

The fact is that every piece of Florida's natural environment is precious. Perhaps John Muir said it best: "When we try to pick out anything by itself, we find it hitched to everything else in the universe."

Armed with this practical, basic book, you can help protect our beautiful state. It contains seven chapters on environmental opportunities, a final chapter on forming groups or lobbying and an appendix of environmental contacts with addresses and phone numbers.

IRREPLACEABLE LANDSCAPES

Since 1950, Florida has lost over half of its wetlands, one-quarter of its forests, and most of its tropical hardwood hammocks, scrub, and coastal habitat. Habitat destruction is linked to almost every environmental problem we have in Florida, from endangered species to water and air pollution. In this chapter, you will learn ways you can get involved in the purchase and protection of Florida's irreplaceable landscapes.

PRESERVING ENVIRONMENTALLY SENSITIVE LANDS

- One acre of high-water recharge land can return 420,000 gallons of rain water into groundwater aquifers each year. This supplies all the water needed for five people for an entire year.

- Fifty-six percent of Florida's historic wetlands have been lost to drainage and filling. At the present rate, 3 million acres of wetlands and forest lands will have been destroyed by the year 2020.

- Thirty-two percent of Florida's upland forests have been converted to agricultural or urban uses.

More than 800 people move to Florida every day. New houses, roads, and services for new residents are eating up lands that once served as green space and wildlife habitat.

Florida still has many wildlands, scenic rivers, and irreplaceable lands that can be purchased for long-term protection. But, environmentally sensitive lands are quickly disappearing and their price is increasing.

Federal, state, local, and non-profit groups are targeting key areas for purchase. The state's Conservation and Recreational Lands (CARL) funds, derived from 9.2 percent of the state's documentary stamp tax revenues and $10 million from phosphate severance taxes, have purchased 600,000 acres of environmentally sensitive lands since 1972. If the new Preservation 2000 program were fully funded, it would greatly expand this successful effort.

Local governments have collected land acquisition funds from bond issues, local option sales taxes, impact fees, and other such innovative sources. The state tends to purchase lands where there is strong local support and matching funds.

Look around and help pinpoint the most vital natural areas in your community. Then work to make sure they are protected for future generations.

SUCCESS STORIES

- As of 1991, 14 Florida counties have approved over $600 million in public funds to acquire and protect lands for conservation, recreation, and open space. Many more are working on land acquisition programs.

- Volusia County's land acquisition program, one of the state's model programs, has purchased 35,000 acres of environmentally sensitive lands.

- The 1990 Florida legislature created Preservation 2000, a $3-billion program to purchase environmentally sensitive lands. The funds are to be divided between the Conservation and Recreation Lands Program (CARL), Save Our Rivers, state parks, Florida Communities Trust, Rails to Trails, forestry lands and Florida Game and Fresh Water Fish Commission acquisitions. However, the funding for the program ($300 million a year for 10 years) has not been permanently established.

- The Nature Conservancy, a national non-profit organization with offices in Florida, has protected more than 400,000 acres of land in the state. Another non-profit group, Trust for Public Land, has protected more than 78,000 acres of land in Florida.

WHAT TO DO

- Purchase and study the *Florida Land Acquisition Handbook* to familiarize yourself with land purchase programs (See resources).

- If your county has not already passed a referendum to establish funds for the purchase of environmentally sensitive lands, set up a group to start the process (See Chapter 8). Use successful counties such as Volusia and Palm Beach for models.

- Identify valuable habitats in your area with the help of the Audubon Society, Sierra Club, the Wildlife Federation, Nature Conservancy and the Florida Native Plant Society. The Florida Natural Areas Inventory, sponsored by the Nature Conservancy, lists areas that are reported to be environmentally sensitive. For information call (904) 224-8207.

- See if lands in your area are already on any of lists to be purchased such as the CARL List or Rails to Trails, etc. (See the *Florida Land Acquisition Handbook* for a complete list.) Recruit local residents to speak at their statewide meetings on behalf of your local project.

- Research the more than 14 land trusts that have been established throughout Florida through the Florida Land Trust Association or the Land Trust Alliance. Determine if your county would benefit from a similar group.

- Volunteer to assist with management of a nearby natural area under national, state, local or Nature Conservancy jurisdiction. Once properties are bought, it is important to make sure that they are protected from trash dumpers, wildlife poachers and arsonists.

- If you own forest lands, learn about the Forest Stewardship Program. Free expert assistance is available to help private forest land owners manage their lands to enhance wildlife, timber, water, and recreation. Call the Florida Division of Forestry, (904) 488-9829, or your county cooperative extension agent under the county listings in the phone directory.

- Support non-profit groups such as The Nature Conservancy which purchase environmentally sensitive lands in Florida.

RESOURCES

THE FLORIDA LAND ACQUISITION HANDBOOK
   by David Gluckman
   Trust for Public Land
   1310 Thomasville Road
   Tallahassee, FL 32303
   (904) 222-9280
   Describes how to obtain state funding for purchase of conservation and recreation lands through a variety of state agencies and water management districts. Cost: $10.

LAND PRESERVATION FOR FLORIDIANS
   Florida Land Trust Association
   1105 North Main St., Box 28-D
   Gainesville, FL 32601
   (904) 375-4560
   A free booklet and companion video available for groups interested in Florida's land acquisition programs.

VOLUSIA COUNTY ENDANGERED LANDS
   ACQUISITION PROGRAM
   123 West Indiana Ave.
   Deland, FL 32720
   (904) 427-2211
   One of Florida's model land acquisition programs.

CONSERVATION AND RECREATIONAL LANDS PROGRAM
   3900 Commonwealth Blvd.
   Mail Station #45
   Tallahassee, FL 32399-3000
   (904) 487-1750
   A major funding source for acquisition of large parcels of conservation and recreation lands.

STATE PARK VOLUNTEER PROGRAM
   Bureau of Park Planning
   Division of Parks and Recreation
   3900 Commonwealth Blvd. MS 525
   Tallahassee, Florida 32399
   (904) 488-8243
   Volunteers perform various services, such as serving as "campground hosts" aiding campground managers with a camping area, greeting visitors, leading tours, removing exotic plants, conducting research or building trails.

LAND AND WATER CONSERVATION FUND (LWCF) AND FLORIDA RECREATION
   DEVELOPMENT ASSISTANCE PROGRAM (FRDAP)
   Department of Natural Resources
   Bureau of Local Recreation Services
   3900 Commonwealth Blvd. Mail Station 585
   Tallahassee, FL 32399-3000
   (904) 488-7896

Competitive matching grant through governments or conservation organizations for the acquisition and development of outdoor recreation projects.

FLORIDA COMMUNITIES TRUST
Department of Community Affairs
2740 Centerview Drive
Tallahassee, FL 32399-2100
(904) 922-2207
Land acquisition funds available to cities and counties to help implement Conservation, Recreation, Open Space, and Coastal elements of local comprehensive plans.

THE NATURE CONSERVANCY
Florida Regional Office
2699 Lee Road, Suite 500
Winter Park, FL 32789
(407) 628-5887
A strong advocate for the Preservation 2000 program and environmentally sensitive land acquisition in Florida. Ask for their "Preservation 2000 Annual Report."

THE FLORIDA NATURAL AREAS INVENTORY
1018 Thomasville Rd., Suite 200-C
Tallahassee, FL 32303
(904) 224-8207
An inventory of environmentally sensitive lands throughout Florida.

COMMON GROUND NEWSLETTER
The Conservation Foundation
1800 N. Kent St., Suite 1120
Arlington, VA 22209
(505) 984-2871
A great source of information about land and water conservation in America.

STARTING A LAND TRUST
Land Trust Alliance
900 17th St. NW, Suite 410
Washington, DC 20006-2596
(202) 785-1410
Order this complete book about forming a local land trust or ask for their free booklet. Book cost: $12.

PRESERVING FAMILY LANDS (FL)
P.O. Box 2242
Boston, MA 02107
A good source of basic information for landowners about taxes and conservation issues. Cost: $6.

"GUIDE TO NATURAL COMMUNITIES OF FLORIDA"
BROCHURE
Florida Department of Natural Resources
Office of Land Use Planning
3900 Commonwealth Blvd.
Tallahassee, FL 32399
(904) 487-1750
A quick overview of the various ecosystems in Florida.

ECOSYSTEMS OF FLORIDA
University Presses of Florida, Orlando
R.L Myers and J.J. Ewel
An extensive guide to Florida's landscape: its forests, fresh waters, marshes, and marine life.

WHY WASTE A WETLAND

- The 1984 Henderson Wetlands Act requires the Florida Department of Environmental Regulation to consider "mitigation" of wetlands. Developers or companies can agree to create man-made wetlands to substitute for loss of true wetlands.

- Although many companies have tried very hard to create ecologically sound wetlands, of the 119 created wetland sites reviewed from 63 Florida permits, only four were in complete compliance.

- According to a recent Florida Department of Environmental Regulation study, the ecological success for created wetlands is only 27 percent. (Freshwater sites are less successful than tidal wetlands.)

- According to the *Miami Herald* (May 24, 1991), only 3 percent of all permits to develop wetlands are denied each year.

- A study of wetlands indicated that each acre produces between $2,300 and $9,800 in food annually. For example, the wetland areas around Apalachicola Bay were valued at $3.8 million in 1987. Acre for acre, they are more productive than agricultural lands.

- Mangrove swamps and tidal areas are nurseries for two-thirds of all fish caught worldwide. In fact, mangroves are so valuable that the city of Miami fines up to $25,000 per tree if mangroves are illegally removed.

- Seventy-five percent of commercial fish and shellfish in the United States are species that depend on coastal wetlands and estuaries during some phase of their life cycles.

Wetlands are areas such as swamps, hammocks, marshes, and prairies that are frequently covered with water. Between 1850 and 1973, Florida lost 12 million acres or 56 percent of its wetlands. Only 11 million acres remain. Wetlands are the most valuable of the Florida ecosystems, environmentally and economically.

Why are wetlands so valuable? In addition to being a nursery and spawning grounds for fish and shellfish, they provide a place to store rain water, a natural filtering system to clean water, and a trap for petroleum wastes, sediments, pesticides, and heavy metals.

Wetlands also help control flooding and erosion. They provide nesting areas for waterfowl and habitats for approximately 50 percent of Florida's endangered species. Scientists believe that the disappearance of wetlands has affected Florida's rainfall. Central and south Florida have experienced a 10 percent decline in rainfall which could be linked to the fact there are fewer wetlands for sources of evaporated water. Some of

Florida's most famous wetlands are the Everglades and the Big Cypress Swamp, coastal mangroves, and the Green Swamp of Central Florida.

SUCCESS STORIES

- St. John's River Water Management District has started a five-year project to restore almost 40,000 acres of diked and stagnant salt marshes along the Indian River Lagoon to a healthy breeding ground for a variety of aquatic animals. Water quality is improving and fish populations are thriving. St. John's River Water Management estimates that the project will benefit commercial and sport fisheries by up to $50 million.

WHAT TO DO

- Don't dredge and fill or alter even small wetlands on your personal property.

- On a lakeside lot, build longer docks to bypass grassy areas.

- Use vegetated rip-rap (sloped stone fill) instead of vertical sea walls.

- Encourage local governments to build bridges instead of "fill" causeways.

- Discourage channeling rivers or using water management systems that affect the natural flow of water.

- Support the purchase of wetlands by local and state agencies and non-profit groups.

- Report wetland destroyers to the Department of Environmental Regulation and support the tightening of regulations and the imposition of stiff fines.

- Large companies are often allowed to destroy wetlands if they agree to create an artificial wetland to replace the loss (called "mitigation"). The result is often an inferior wetland. Lobby for less mitigation, particularly where other alternatives exist. Consider a mitigation bank or funds from companies destroying wetlands that will be used to purchase other natural wetlands for preservation.

- Encourage landowners to maintain their wetlands without losing economic benefits from the land by waterfowl, shellfish, and marine fish.

RESOURCES

FLORIDA DEPARTMENT OF ENVIRONMENTAL REGULATION
The Bureau of Wetland Resource Management
2600 Blair Stone Road
Twin Towers Office Building
Tallahassee, FL 32399-2400
(904) 488-0890

Free booklet on *State of the Wetlands*

THE NATIONAL ENVIRONMENTAL PROTECTION AGENCY
Region IV
345 Courtland St.
Atlanta, GA 30308
(404) 347-4728 or 800-832-7828

Information on national wetland protection regulations and pending legislation.

"OUR VITAL WETLANDS" Booklet and Brochure
Lake County Water Authority
107 North Lake Avenue
Tavares, FL 32778
(904) 343-3777

Clearly describes the importance of wetlands.

FLORIDA DEFENDERS OF THE ENVIRONMENT *MONITOR* NEWSLETTER
May-June 1991
Mitigation article by Kerry Dressler
1523 N.W. 4th St.
Gainesville, FL 32601
(904) 372-6965

Legislative recommendations for the regulation of wetlands.

STREAMLINES NEWSLETTER
ST. JOHN'S RIVER WATER MANAGEMENT DISTRICT
P.O. Box 1429
Palatka, FL 32178-1429
(904) 329-4500

Ask for a copy of the June/August 1991 edition of *Streamlines* featuring the Indian River Lagoon Project.

FLORIDA'S GREATEST WETLAND: THE EVERGLADES

The Florida environment could be compared to the human body in many ways. If you put enormous stress on one part of the body, it affects the functions of several others. Like the human anatomy, the environment is one complete and interlocking system where everything is connected. One of the greatest examples of such a system is the Florida Everglades, a 1.4-million acre marsh. The Everglades National Park is one of the most threatened national parks in the United States. Some pessimists project that the Everglades' illness is terminal. They claim it will die within 10 years.

If the Everglades died, millions of residents and a wide diversity of wildlife would be endangered.

The Everglades system, which includes the Kissimmee River and Lake Okeechobee, is the key to South Florida's water supply. From Miami to the Keys and parts throughout South Florida, the Everglades keeps the salt of seawaters from intruding into the freshwater drinking supply. Once the waters reach the Everglades, there are millions of south Florida residents and expansive farms ready to suck down the water table. South Florida residents use an average 200 gallons of water a day, compared to the national average of 105 gallons per day. Additionally, the Everglades Agricultural Area boasts 450,000 acres of thirsty sugarcane and vegetables.

The enormous water system begins with a chain of lakes south of Orlando, flows south to the Kissimmee River and Lake Okeechobee, and finally spills over into the Everglades. Whatever happens to the water along the way affects the life expectancy of this great river of grass.

The Everglades and connecting lakes and rivers have suffered a tremendous decline in water quality. Like all wetlands, the Everglades was meant to be a huge natural filtering system to soak up nutrients and flush out pollution. Unnatural demands from excessive pollution have overloaded the delicate region. In addition, this broad, shallow, life-giving "river of grass" has been dredged, dammed, diked, and abused through years of misunderstandings. In the early years, speculators saw the Everglades as a worthless swamp, useful only if it could be dredged and filled.

Contaminated runoff into Lake Okeechobee contains 10 to 20 times normal concentrations of phosphorus and nitrogen. The dairy farms of Okeechobee produce as much waste as a city of 98,000 persons. For years, the sugarcane farms have polluted the Everglades with runoff containing excess nutrients and pollutants from fertilizers and pesticides.

In addition to its need as a vital link in South Florida's tenuous water crisis, the Everglades has enormous wildlife diversity. More than a dozen endangered or threatened species including the Florida panther, snail kite, wood stork, American crocodile, and five species of sea turtles inhabit the park. The huge wetland is home to more than 330 species of birds. However, 90 percent of the park's wading birds have disappeared in the last 50 years due to the destruction of nesting areas. Sawgrass, the mainstay of the park and one of the oldest plants on earth, is being threatened. Nutrient-loving cattails are invading the marshes. The plants grow to over 10 feet and can block out sunlight to many species that are important to the food chain.

Today, the Everglades, the third largest subtropical wilderness in the United States, has been recognized as an International Biosphere Reserve by the United Nations and is a World Heritage Site. But the future of the river of grass depends on statewide cooperation from individuals, agricultural interests, politicians, and environmental groups. The move to correct past problems has begun. Awareness of the Everglades' environmental problems is increasing. Attitudes are changing. The government, dairy farmers and sugar cane growers must work together to save the Everglades. Environmentalists hope it is not too late.

RESOURCES:

THE EVERGLADES PROTECTION PROGRAM
Executive Office of the Governor
The Capitol
Tallahassee, FL 32399-0001
(904) 488-5551

EVERGLADES COORDINATING COUNCIL
610 Northwest 93 Avenue
Pembroke Pines, FL 33024
(305) 653-1416

EVERGLADES PROTECTION ASSOCIATION
P.O. Box 216
Islamorada, FL 32901

FRIEND OF THE EVERGLADES
3744 Steward Avenue
Miami, FL 33130

EVERGLADES COALITION
900 17th St.,NW
Washington, DC 20006
(202) 833-2300

VOICE OF THE RIVER and THE EVERGLADES: RIVER OF GRASS
by Marjory Stoneman Douglas

GET SET FOR GREENWAYS

- Wildlife corridors are essential for species' survival. Endangered Florida panthers roam over 200 miles a day. Florida black bear travel as far as 20 miles a day. Even the otter needs 10 miles of uninterrupted river bank to travel along each day.

- The United States has more than 150,000 miles of abandoned rail lines available for conversion to trails through the "Rails to Trails" program. Thirty-five states have converted more than 3,000 miles to trails.

- The American Gas Association and the Conservation Fund are exploring whether the one million miles of U.S. national gas pipelines can be used for right-of-ways for greenways.

- More than 100 greenway projects are in progress nationally. The Conservation Fund has a greenway database with details on each project.

- A statewide initiative to create a complete greenways network is now underway through the 1000 Friends of Florida and The Conservation Fund. Florida already has the 1,000-mile Florida Trail that spans from the panhandle to South Florida.

- One mile of rail corridor represents 12.5 acres of greenway. Currently $3.9-million from Florida's Preservation 2000 Fund is earmarked for Rails-to-Trails lands. More than 100 miles of right-of-way have been purchased since 1987 and the purchase of 78 more miles is underway.

Greenways are natural walkways or green corridors for people and wildlife. These corridors of protected open spaces edging streams, ridge lines, connecting parks, refuges, and historic sites provide outdoor recreational opportunities, important habitats, wildlife corridors, and a network of open space. Urban greenways break up the urban sprawl and link communities.

The word "greenway" was coined by combining the words "greenbelt" and "parkway." Wisconsin landscape architect Phil Lewis dubbed them "E" ways for Environment, Ecology, Education, and Exercise.

Wildlife needs networks of green space. Shrinking habitat causes inbreeding and leads to extinction. Small, stand-alone parks do not link various natural spaces for animals that need broad spaces to breed and roam. For instance, linking several national or state parks through a narrower greenway may provide the extra habitat necessary for the Florida panther and the black bear. Corridors also are good for plants and animals because wild roaming animals disperse seeds throughout the corridor system.

However, some environmentalists are skeptical that wildlife will use greenways.

Still, the Nature Conservancy advocates the purchase of remnant habitats wherever possible. Where larger tracts of natural habitat can be purchased, they should be. Where large land holdings are not available, greenways can serve a valuable purpose.

Greenways offer more than merely a solution to environmental problems. They touch on wellness and social issues, too. People have a chance to walk, bike or run in the open spaces, away from the stresses of urban life. A very successful greenways system in Austin, Texas, provides 1,800 acres of greenbelts and linear parks running beside Austin's 18 creeks. The oasis of urban green space contains 25 miles of walking and biking trails and protects Austin's valuable streams.

SUCCESS STORIES

- The 30,000-acre Pinhook Swamp in north Florida is one of the largest wildlife corridors in North America. The 10-mile-long corridor connects two major rivers and provides an "auto-free" expressway for Florida black bears traveling from the Okefenokee National Wildlife Refuge to the Osceola National Forest. The Nature Conservancy has played a vital role in protecting this nearly unbroken swamp/river ecosystem.

- Florida Trail—The Florida Trail Association, founded more than two decades ago, has helped develop more than 1,000 miles of prime hiking trails and sponsors about 500 events annually. However, even the Florida Trail is being threatened by development. For the second year, it has been listed as one of the most endangered trails by the American Hiking Society.

- State officials are laying the groundwork for a linear 73,000-acre park from former Cross Florida Barge Canal lands now called the Cross Florida Greenbelt area.

FUNDING IDEAS

- High Point, North Carolina—Residents sold deeds to foot-long sections of their greenway to raise money for its development.

- Pueblo, Colorado—Builders created a centerline through the greenway of bricks with donors names etched on them.

- Oregon—Residents developed the 40-Mile Loop Land Trust. Volunteers scouted routes, talked to landowners, got land donations, marked routes and mapped the system. The city's summer youth program provided the building power to pave the trail.

WHAT TO DO

- Create your own mini-corridor along your fence line, lake or riverfront lot, or other area where you can connect natural wooded areas by letting the area grow back naturally.

- Find a local group or create a group to work on a greenways project in your community.

- Identify key areas for greenways with the help of a landscape architect or planner and biologist.

- Search for innovative, inexpensive greenways such as gas pipeline right-of-ways and abandoned railroad tracks.

- Develop a master plan for a local greenway system and work towards obtaining grants and private contributions of land and money to complete the project. Research done in Seattle reveals that property sells faster when it faces a recreation corridor.

- Use technical assistance provided through the National Park Service and The Florida Greenways Project.

- Join the Florida Trail Association and help complete 1,300 miles of preserved trails and support a great educational resource.

- Write the U.S. Forest Service to support funding of diminishing forest trails rather than build unnecessary forest roads, which often destroy trails.

RESOURCES

THE FLORIDA GREENWAYS PROJECT
1000 Friends of Florida/The Conservation Fund
P.O. Box 5948
Tallahassee, FL 32314-5948
(904) 222-6277
Ask for ways you can help support the project locally.

FLORIDA TRAIL ASSOCIATION
P.O. Box 13708
Gainesville, FL 32604
Toll Free 800-343-1882
Ask for a free brochure about the trail, membership dues, and volunteer opportunities.

FLORIDA RAILS-TO-TRAILS CONSERVANCY
2545 Blairstone Pines Drive
Tallahassee, FL 32314
(904) 942-2379
Obtain information on funding for local greenways.

GREENWAY FACT SHEETS
Scenic Hudson
9 Vassar St.
Poughkeepsie, NY 12601
(914) 473-4440

Booklets covering greenway design, construction, volunteer recruitment, fund-raising, land trusts, etc. Cost: $8.

ECONOMIC IMPACTS OF PROTECTING RIVERS, TRAILS, AND GREENWAY CORRIDORS
National Park Service 1990
P.O. Box 37127
Washington, DC 20013

Greenways create positive economic benefits, such as increased property values. This resource book will help community groups better communicate the benefits of greenways.

GREENWAYS FOR AMERICA
by Charles E. Little
A current book on the greenways movement in America. Cost: $22.95.

CONSERVATION FUND
1800 N. Kent St., Suite 1120
Arlington, VA 22209
(703) 525-6300

Order their newsletter to stay up-to-date on greenway projects around the nation. Ask for information on utilizing gas pipeline right-of-ways for greenways. *Greenways Design Manual* will be published in 1992.

THE NATURE CONSERVANCY
Florida Regional Office
2699 Lee Road, Suite 500
Winter Park, FL 32789
(407) 628-5887

Get more information about the Pinhook Swamp project and other similar projects in Florida.

DEVELOP DIFFERENTLY

- In 1950, suburbanites made up a third of the United States' population. Today, suburban dwellers are in the majority.

- American Farmland Trust estimates that every hour 320 acres of agricultural land is lost to development in the United States.

- Clustering planned-unit development reduces the cost of development by an average of $4,000 to $7,000 per dwelling unit. It cuts the costs of sewer lines, roads, and other services.

- Automobiles consume a tremendous amount of urban space. In Los Angeles, two-thirds of downtown is used for driving, servicing, and parking cars.

For more than 40 years, Americans have been in love with the idea of the large-scale suburban housing developments. But the car-dependent suburb has damaged the environment and diminished America's sense of community.

Urban sprawl invades land important for environmental and natural resource protection and increases the cost for governmental services. Roads, school transportation, sewer and water pipes, and fire and safety coverage cost more when they are stretched throughout a county. Unfortunately, conventional zoning codes have promoted "planned sprawl."

Recently, architects—such as Miami's Andres Duany and his wife Elizabeth Plater-Zyberk—have returned to planning small towns and neighborhoods. A traditional neighborhood design calls for smaller front yards and narrower roads to bring neighbors together. Plans often include common open spaces and mixed-use development so people can walk or ride bikes to work, shopping, and recreational areas.

Surprisingly, all the research points out that the property in Duany-designed towns such as Seaside, Florida, and downtown Starke goes up in price because this is attractive and very liveable.

Similar concepts have been used in the New England states to concentrate growth within towns and protect surrounding farmland. Use of open-space zoning allows the same overall amount of development on (typically) half the parcel. Developers look at their land first to select the most unique and attractive open spaces; then they plan the development in the less environmentally sensitive places.

SUCCESS STORIES

- Seaside, a mixed-use resort on Florida's Panhandle, has been so successful that initial property prices have increased 500 percent.

- The Village of Tannin, on the Gulf Coast of Alabama, was carefully designed to ensure trees were not lost. Houses were moved four feet back in one section just to save a row of oak trees. Curbs were constructed with a specially designed machine to prevent disturbing trees beyond 12 inches from the roadway.

WHAT TO DO

- Persuade local zoning boards to revise their codes to allow for traditional neighborhood designs. Duany developed a set of sample ordinances available called "Traditional Neighborhood Development Ordinances" or "TNDs" that were cited by the Florida Governor's Task Force on Urban Growth Patterns as an outstanding model.

- Become familiar with traditional neighborhood design and encourage local builders and developers to consider using some of the techniques. The resource list contains several current articles on traditional neighborhood design.

- Encourage city and county governments to plan for open space and discourage sprawl through well-developed comprehensive plans.

RESOURCES

DEPARTMENT OF COMMUNITY AFFAIRS
2740 Centerview Drive
Tallahassee, FL 32399-2100
(904) 488-4545

Request copies of DCA's *Technical Memos* related to traditional neighborhood design and planning for rural lands.

HOME BUILDER DESIGN AND EDUCATION PROGRAM
Governor's Energy Office
The Capitol
Tallahassee, FL 32399-0001
(904) 488-2475

State program to improve housing and development planning to conserve energy, land and water resources. Conducts workshops for developers and planners.

AMERICAN FARMLAND TRUST
1920 N Street, N.W., Suite 400
Washington, D.C. 20036
(202) 659-5170

AFT has a variety of studies and brochures on preservation of farmland, preserving open space and other related topics.

CENTER FOR RURAL MASSACHUSETTS
University of Massachusetts
Amherst, MA 01003
(413) 545-2255

Distributes a video on traditional neighborhood design, various studies, and publications.

ANDRES DUANY AND ELIZABETH PLATER-ZYBERK
Architects and Town Planners
1023 S.W. 25th Avenue
Miami, FL 33135
(305) 644-1023

Ask for information on traditional neighborhood design and sample ordinances.

DESIGN WITH NATURE
by Ian McHarg

Landscape architect and prime mover in the landscape architecture/ecology movement explains how to design urban areas that do not intrude on nature.

THE GRANITE GARDEN
by Anne Whiston Spirn

A book about urban nature and human design.

PROMOTE GROWTH MANAGEMENT

- Florida's population is projected to rise to 20 million by year 2002.

- If tourism holds steady, 400 million people will visit Florida during the next 10 years.

Effective growth management at the local and state levels can greatly improve the quality of life, protect natural and cultural resources, and enhance economic opportunities. As of August 1, 1991, every local government in Florida had submitted a draft comprehensive plan to the Department of Community Affairs. The 459 plans completed the first cycle of Florida's Growth Management Laws that were adopted in 1985. This legislation limits urban sprawl and directs county governments about appropriate land uses. New development plans must provide for drinking water supplies, waste disposal, roads, schools and other vital services.

Growth-management followers predict that in the next couple of years, the laws will be reviewed and revised. Growth-management opponents are trying to eliminate the legislation or to water it down.

WHAT TO DO

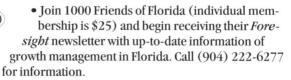

- Join 1000 Friends of Florida (individual membership is $25) and begin receiving their *Foresight* newsletter with up-to-date information of growth management in Florida. Call (904) 222-6277 for information.

- Contact your local county and city governments and find out where they are in the comprehensive planning process. Are they developing regulations to support the plan? Are they implementing the plan? Buy a copy of the goals, policies and objectives of the local comprehensive plan to track whether the local government is doing what it planned.

- Plan amendments can be made every six months. Be on the lookout for unnecessary plan amendments. Meeting notices regarding growth management meetings must be placed in your local newspaper.

- Study model communities to gather input for ways to make your community more livable.
- Support growth management legislation through letters to your state and local officials.

RESOURCES:

CITIZEN'S HANDBOOK TO THE LOCAL GOVERNMENT
COMPREHENSIVE PLANNING ACT
by Casey and David Gluckman
Florida Audubon Society
460 Highway 436, Suite 200
Casselberry, FL 32707
(407) 260-8300

A detailed guide outlining exactly how you can get involved in the comprehensive plan process. Contains sample letters, flow charts and key legislative information.

1000 FRIENDS OF FLORIDA
Post Office Box 5948
Tallahassee, FL 32314-5948
(904) 222-6277

Provides support to residents involved in county comprehensive plans. Ask for information on *Growth Management 2000: Making it Work*, a recently released book.

DEPARTMENT OF COMMUNITY AFFAIRS
2740 Centerview Drive
Tallahassee, FL 32399-2100
(904) 488-4545

DCA reviews all comprehensive plans and determines whether they comply with Florida growth-management legislation. Request copies of reports written about your local comprehensive plan, the *Elms II* report and the *Urban Sprawl Task Force* study.

CREATING SUCCESSFUL COMMUNITIES:
A GUIDEBOOK TO GROWTH
MANAGEMENT STRATEGIES
P.O. Box 7
Covelo, CA 95428-9901
800-828-1302

Workable action plans for creating a vital and livable community. Strategies for participation in the growth management planning process. Cost: $26.95

LOCAL COUNTY AND CITY PLANNING OFFICES

RECYCLE A HOUSE OR OFFICE

- Because of the glut of commercial space in our cities, experts project that 80 percent of commercial development in the 1990s will be retrofitted existing structures.

When asked what single most important thing individuals could do to protect Florida's environment, Florida Audubon Society Director Charles Lee answered, "Buy an existing house or a lot in an already established neighborhood." Each new house takes away a piece of wildlife habitat and green space.

Before building a new house or office, Lee recommends looking at the vast numbers of empty houses, lots, and offices around your community. The construction boom of the 80s left Florida with a surplus of vacant buildings.

SUCCESS STORIES

- In Miami, part of the Lincoln Road Mall and the old Merita bread factory were converted into arts cooperatives.

- An old school for wayward girls, The McPherson School for Girls, has been converted into county offices in Marion County.

- An almost vacant mall in Tallahassee, Northwood Mall, has been utilized for state agency offices.

- The historic Marion Theatre in downtown Ocala, Florida, is being converted into a hands-on children's science museum.

WHAT TO DO

- Get the facts on how much vacant office, commercial and residential space is available in your community. See your city or county comprehensive plan for figures in your area.

- Pinpoint blighted, vacant areas and interesting, unique historic buildings to restore and redevelop.

- Encourage the city and county to become leaders in utilizing existing spaces before building new structures for offices.

- Establish incentives for businesses utilizing existing spaces such as faster permit issuance, lower impact fees, or a community award program.

RESOURCES

FLORIDA TRUST FOR HISTORIC PRESERVATION
P.O. Box 11206
Tallahassee, FL
(904) 224-8128

LOCAL CITY OR COUNTY PLANNING OFFICES

LOCAL HISTORIC PRESERVATION SOCIETY

KEEP FLORIDA FIRST IN TREES

- Urban areas without trees are 3 to 5 degrees hotter than their surroundings.

- Three properly placed trees can cut air-conditioning bills by 5 to 30 percent.

- The greenhouse effect can be reduced by planting more trees, especially within urban areas. Urban trees planted around parking lots and other "heat islands" are 15 times more important in reducing carbon dioxide build-up than rural trees.

- One forest tree absorbs 26 pounds of carbon dioxide a year and releases oxygen.

CO_2 → 🌲 → Oxygen

- According to a U.S. Forest Service study, trees add 5 to 10 percent or approximately $3,000 to $7,000 to the value of a home.

- A mill designed to use waste paper instead of wood pulp is 50 to 80 percent cheaper to build.

Did you know that Florida is number one in the nation for diversity of trees? Florida has more than 314 species and 26 million acres of state and privately owned forest and wildland.

However, Florida has lost three million acres of forest lands since 1930 and at the present rate will lose another three million acres of forest and wetlands by the year 2020.

For every tree that is planted, four die. Trees are destroyed from construction damage, improper care and selection, misuse of herbicides, vandalism, and automobile accidents.

SUCCESS STORIES

- Dade County Schools—After 1990, all public schools built in Dade County must provide the maximum tree canopy possible after providing open space mandated for school-related projects. Trees are especially desired over "heat islands" or large expanses of concrete or asphalt.

The new program was initiated to reduce heat gain in the schools, limit the power needed to cool schools, and assist in deterring global warming. In most counties, school systems are major landholders. Just think of the impact if all county schools and public properties practiced this same policy!

- "Our Forest, Their Forest", Children's Program in Miami, Florida—elementary schools plant their own mini-forests on the school grounds.

- Gainesville, Florida, is both a city and a forest. According to a 1984 analysis, 46 percent of the city was shaded. Gainesville's parks division still is planting trees. It planted 400 between 1989 and 1990.

WHAT TO DO

WHAT TO DO

- Make sure your community has a "tree board," appointed by the city or county government to make sure trees are maintained and/or replanted in your community.

- Search out funding for urban forestry programs from private and public sources.

- Donate money from recycling campaigns to community tree programs.

- Develop a community newsletter with lists and locations of trees and plants needing new homes. (Houston, Texas—*Operation Plant Rescue* and Seattle, Washington— *Plant Amnesty*) Developers who remove trees and plants can offer them for free to a good home.

- Encourage city and county governments to allocate adequate money for tree upkeep and maintenance.

- Adopt local tree ordinances and scenic road policies. Gainesville has a model tree ordinance and Tallahassee, Florida has a good scenic road policy.

- Obtain help from the state forester, county cooperative extension agent, or local nursery owner.

- Work towards a "Tree City USA" designation.

- Plant non-maintenance public-access fruit and nut trees in low-income sections of your community. (Fruition Project—Dade County, Florida.)

RESOURCES

COUNTY COOPERATIVE EXTENSION AGENT

LOCAL FORESTER

TREE CITY USA PROGRAM
National Arbor Day Foundation
100 Arbor Ave.
Nebraska City, NE 68410
(402) 474-5655
Assists cities in obtaining Tree City USA designation.

PLANT A TREE FOR LIFE PROGRAM FOR FIFTH GRADERS

Florida Dept. of Agricultural and Consumer Services
3125 Conner Blvd.
Tallahassee, FL 32399-1650
(904) 488-7617

GLOBAL RELEAF
American Forestry Assoc.
P.O. Box 2000, Dept. FL
Washington, D.C. 20013
(202) 667-3300

FRUITION PROJECT and OUR FOREST, THEIR FOREST
Urban Forester
Metro-Dade County
111 N.W. 1st St., Suite 1310
Miami, FL 33128
(305) 375-3376

OPERATION PLANT RESCUE
Houston Parks Department
Houston, TX
(713) 845-1000

PLANT AMNESTY
Seattle Information Center
Seattle, WA
(206) 684-4557

DADE COUNTY SCHOOLS TREE CANOPY PROGRAM
Dade County Landscape Planning
Miami, FL
(305) 375-2810

GAINESVILLE TREE PLANTING PROGRAM
Parks Division
P.O. Box 490 Station 27
Gainesville, FL 32602
(904) 334-2171

TALLAHASSEE/LEON COUNTY SCENIC ROADS
Canopy Road Program—Leon County Planning
300 S. Adams St.
Tallahassee, FL 32301
(904) 599-8600

A WILDLIFE WONDERLAND

Florida is blessed with an enormous variety of plant and animal life. In fact, among all the states, only Texas and California have greater wildlife diversity.

But Florida's future as a wildlife wonderland is threatened. Since 1950 we've lost almost half of our wetlands to development, and paved more than a quarter of our forested acres. Most of our tropical hardwood hammocks, scrub, and coastal habitats have also been lost. Introduced aggressive plants have taken over native Florida plant communities in many parts of the state, including state and national parks.

As their unique ecosystems are degraded or destroyed, roughly half of Florida's vertebrate species are declining. In this chapter you will learn ways you can help restore and preserve critical habitats for Florida's animal and plant species.

LANDSCAPE FOR WILDLIFE

We humans have a habit of sub-dividing animal habitats into building lots, and forgetting the needs of the original inhabitants when we landscape. With a little effort, we can entice many native species to return.

This is becoming increasingly important as growth consumes more and more of Florida's green space. It's no longer enough to set aside forests and wilderness areas for wildlife. Our city and county parks, industrial facilities, executive parks, cemeteries, churchyards and road right-of-ways, as well as residential subdivisions and individual homesites offer potentially valuable wildlife habitats. Even city balconies, rooftop gardens and window boxes can provide homes for some of Florida's smaller critters.

In planning your landscape for wildlife, keep in mind the four basic needs common to all creatures: food, water, cover, and space. Of these, water is most frequently neglected. The lack of dependable water sources often keeps wildlife from otherwise suitable suburban and urban areas.

WHAT TO DO

• Learn all you can about natural areas near your home, and plan your landscape to be as much like these natural habitats as possible.

• Choose native plant varieties which will supply year- round food. Plants bearing early summer berries (such as blackberry and blueberry) and those with fruits lasting through the winter and spring (such as sweetgum, juniper, and holly) will satisfy the food preferences of many species.

• Group smaller trees, shrubs and groundcovers around the taller trees to create an understory. Aim for diversity of heights and densities to satisfy the nesting and cover requirements of various animals.

• Provide fresh water. A birdbath will do if space is limited, but a small pond is better. (Check your local ordinances and laws pertaining to ponds.)

• Learn to appreciate the seasonal flooding that plays a critical part in the life cycles of many Florida species. If you live on acreage with low wet spots, don't fill them in. You'll be destroying valuable habitat.

- Leave dead trees standing whenever possible to provide habitat for cavity-nesting birds and mammals. Substitute man-made nest boxes if dead trees must be removed.

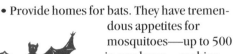

A Wildlife Wonderland

- Provide homes for bats. They have tremendous appetites for mosquitoes—up to 500 in one hour—making them welcome residents in Florida neighborhoods.

- Mulch beds of trees and shrubs with natural materials such as lawn clippings or leaf litter to hold moisture.

- Avoid using pesticides that may kill the wildlife you hope to attract. See "Protect People, Pets, and Wildlife from Poisons," page 42 for alternatives.

- Encourage your highway department to plant shrubs and trees in median strips. In addition to providing habitat for invertebrates and resting places for birds and butterflies, the plants cut the glare of oncoming headlights and serve as barriers that help prevent head-on collisions.

RESOURCES

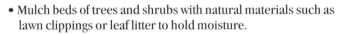

NONGAME WILDLIFE PROGRAM
Florida Game and Fresh Water Fish Commission
620 South Meridian St.
Tallahassee, FL 32301

Offers *Planting a Wildlife Refuge*, a booklet with details on how to create a backyard habitat, and *What Have You Done for Wildlife Lately?; A Citizen's Guide to Helping Florida's Wildlife*. Subscribe free to the newsletter *The Skimmer* for regular news about nongame wildlife. Subscribe to *Florida Wildlife magazine* for $7 per year.

FLORIDA CONSERVATION FOUNDATION
1251-B Miller Ave.
Winter Park, FL 32789
(407) 644-5377

Ask for the brochure *Common Natural Areas of Florida*.

FLORIDA BUTTERFLIES
by Eugene R. Gerberg and Ross H. Arnett, Jr.

NATIONAL INSTITUTE FOR URBAN WILDLIFE
10921 Trotting Ridge Way
Columbia, MD 21044
(301) 596-3311

A nonprofit scientific and educational organization dedicated to the conservation of wildlife and habitat for the benefit of people in cities, suburbs and developing areas. Publishes a quarterly newsletter. Maintains a national registry of urban and suburban wildlife sanctuaries.

NATIONAL WILDLIFE FEDERATION
1400 16th St. N.W.
Washington, D. C. 20036-2266
800-432-6565

Offers *Backyard Wildlife Habitat* pamphlet for $4.95. Ask for item #79919. Maintains a national registry for backyard habitats.

FLORIDA NATIVE PLANT SOCIETY
P. O. Box 680008
Orlando, FL 32868
Publishes a quarterly newsletter for members, and provides information on the protection, management, and restoration of native plant ecosystems.

WHITE'S BIRDBOX CO.
5153 Neff Lake Rd.
Brooksville, FL 34601

This Florida company sells a variety of bird and bat nest boxes at reasonable prices.

MINNESOTA DEPARTMENT OF NATURAL RESOURCES
Box 7
Centennial Building
St. Paul, MN 55155

Ask for their booklet *Woodworking for Wildlife—Homes for Birds and Wildlife*.

SOME WEEDS ARE WONDERFUL

- Today, species are becoming extinct at 25,000 times the natural rate. Every plant, tree, and animal counts.

- The loss of one plant can cause the loss of as many as thirty animals or insects.

- The world produces $40 billion a year in medicines produced from wild products.

- Only a few hundred wildflower species are available in seed catalogs in the United States. Florida native species are very costly and hard to locate.

- Florida has about 40 species of plants on the Federal endangered list.

A Wildlife
Wonderland

Five centuries ago, Ponce de Leon named our state Florida—the land of flowers. He was impressed with the abundance and beauty of Florida's native plants. Today, Florida has more than 3,800 species of plants, both native and those that have been introduced over the decades. Florida's outstanding plant variety is due to our climate, and diversity of geology and geography.

Our roadsides can be corridors for native plants. Planting new wildflowers is good, and managing lands for the expansion of existing plants and wildflowers is even better.

SUCCESS STORIES

- The Florida Department of Transportation planted more than 5,000 pounds of wildflowers in 1991. Seeds are donated from organizations throughout the state. Two good examples include, two hundred miles of crimson clover on I-10 from Tallahassee to Pensacola and 200 miles of phlox along I-75 from Tampa to the Florida border.

- On S.R. 65, the Department of Transportation, working with such groups as the Florida Native Plant Society, has expanded the population of one wildflower—"harpers beauty"—from 70 plants to over 6,000.

- The Sarasota Chapter of the Florida Native Plant Society has adopted a small, neglected park in Sarasota, The Robarts Memorial Mini Park. They planted almost 300 native plants.

• Tallahassee has developed a Gateway Project. The city will plant wildflowers from the airport leading into downtown.

WHAT TO DO

1. The best way to expand native plants and wildflowers is to pinpoint species on public lands and along public roadways. Seek management assistance from your regional Department of Transportation office or parks manager to protect and expand that species.

2. Special mowing techniques should be utilized. When feasible, the area should not be mowed. Otherwise, the area should be mowed after seeds are set and the grass clippings should be scattered in adjacent areas to promote growth.

3. To encourage expansion, grass can be removed around wildflower areas.

4. Shrubs and tree vegetation should be left in a natural state along upper roadway slopes and fence lines to retain a complete ecosystem. Natural vegetation screens often increase property values.

5. If natural wildflowers are not available, work with the Department of Transportation to plant phlox or crimson clover. The seeds must be purchased and given to your regional Department of Transportation office for planting in November and December. Blooms appear in the spring.

6. Encourage your city or county to apply for a Florida Department of Transportation Beautification Grant to plant native plants in selected areas.

RESOURCES

FLORIDA DEPARTMENT OF TRANSPORTATION
Wildflower Project
605 Suwanee St. M537
Tallahassee, FL 32399
(904) 488-2911

Ask for information on how you or your group can donate wildflower seeds. Free posters featuring Florida wildflowers are available as long as supplies last.

FLORIDA DEPARTMENT OF TRANSPORTATION
Beautification Grant
(Same address and phone number as above)

Encourage local governments to apply for the matching grant for beautification of state highways right-of-ways.

LOCAL GARDEN CLUBS

COUNTY COOPERATIVE EXTENSION AGENTS

FLORIDA NATIVE PLANT SOCIETY
P.O. Box 680008
Orlando, FL 32868
(407) 299-1472

Ask for *Planning and Planting a Native Plant Yard*.

ASSOCIATION OF FLORIDA NATIVE NURSERIES
P.O. Box 1045
San Antonio, FL 33576
(904) 588-3687

Ask for nurseries in your area that sell native plants.

FLORIDA WILDFLOWERS AND ROADSIDE PLANTS
by C. Ritchie Bell and Bryan J. Taylor

**A Wildlife
Wonderland**

PROTECT PEOPLE, PETS, AND WILDLIFE FROM POISONS

- Florida's heat and humidity can increase skin absorption of some deadly chemicals. That makes exposure to pesticides much more serious in Florida than the same exposure in a dry, relatively cool state such as Colorado.

- Florida ranks first in pesticide and fungicide use in the country, even though we rank thirty-third in planted crop acreage. Close to one-third of all pesticides, insecticides and herbicides used in the state are applied in urban areas.

- Now banned, the pesticide DDT decimated populations of ospreys, bald eagles and peregrine falcons when the chemical accumulated in the fish and ducks the raptors ate.

- The more effective a pesticide is, the quicker it loses its usefulness. Susceptible insects die, while those resistant to the poison breed and pass on their defenses to their "super insect" offspring.

Unfortunately insects enjoy Florida's balmy climate as much as we do. This may be the only state in which a major newspaper conducts an annual contest to find the world's largest cockroach.

Don't be tempted to rely on strong, persistent pesticides to control pests. There's a reason the labels of many pesticides warn against using them near water supplies, or within reach of children and pets. Conventional insecticides are toxic by design—otherwise they wouldn't kill insects. While it may be true that the recommended amount of one product, such as mosquito repellent, will not cause a problem by itself, in combination with the lawn chemicals you're exposed to on the golf course, or fungicides in the garden, it may add up to more than you can handle safely.

Pets frequently get even heavier doses of toxins. In Florida it's almost certain your dog or cat is either dipped, sprayed or powdered in an effort to control fleas and ticks. If he wanders outside he'll most likely be exposed to lawn chemicals, herbicides, fire-ant poison bait, and so on. He deserves better!

And finally, it doesn't make much sense to cultivate native plants for food and shelter, provide water, and set up feeders, and then spread poison to kill insects, moles, and other small creatures. The wren that builds in the nest box we provide could die from eating the spider we've zapped with insecticide.

There are ways to combat pests without resorting to products that may endanger people, pets, and Florida's wildlife. It takes commitment and patience to find the solutions to your particular pest problems, but the effort is well worth it.

WHAT TO DO
.....IN THE GARDEN AND YARD

A Wildlife
Wonderland

WHAT TO DO

- Try letting nature bring harmful insects under control through natural checks and balances. Most plants can maintain productivity even with a 25 percent loss of foliage from insect damage.

- Help beneficial insects establish themselves in your yard and garden to control harmful insects through predation. Lacewings and lady bugs are among the insect helpers you can order through the mail if unavailable locally. Herbs planted among vegetables attract beneficial insects that prey on pests.

- Practice "companion planting." Separate plants that are susceptible to the same insects and diseases, and intersperse plants that are resistant or even repellent to the pests. For example, African marigolds planted with vegetable crops will repel certain harmful nematodes that attack the roots. Garlic planted near roses helps fight black spot. Strong-scented herbs such as oregano, wormwood, or chives help protect nearby plants from insect pests.

- Try physical barriers to keep insect pests away from plants. Make collars of two-inch-wide strips of light cardboard, taped or stapled to encircle the stem. Place them with one inch below the soil line and one inch above.

- Avoid using persistent or very toxic pesticides such as lindane, diquat, paraquat, and ethion.

- Look for the words "caution," "warning," or "danger" on product labels. "Caution" indicates the lowest level of toxicity; "danger" the highest.

- Substitute insecticidal soap sprays for toxic compounds. They are available commercially, or you can make your own by mixing two tablespoons of dishwashing liquid with two cups of water. (See "Simple Substitutes for Toxic Products," page 46 for another solution.)

- When an insecticide is needed, use one derived from plants or minerals. They are relatively safe for humans, pets, and birds, as well as other wildlife. Examples are Rotenone, Pyrethrum, Sabadilla, and Ryania.

- Pick up fruits and vegetables that have fallen to the ground where they may attract pests. Any diseased plant material should also be removed.

- Pull weeds up by the roots.

WHAT TO DO
. . . ON YOUR PETS

- Try an herbal collar or ointment containing eucalyptus or rosemary.

- Sprinkle brewer's yeast on your dog's food (check with the vet for the amount.)

- Use a flea comb. Its closely set teeth remove fleas and eggs from your pet's hair. A quick dip of the comb into any flea soap or dishwashing solution kills them.

- Treat your pet to a bed filled with cedar chips, a natural insect repellent.

WHAT TO DO
. . . IN THE HOUSE

- Bait roaches with a mixture of half boric acid and half powdered sugar. While not as dangerous as persistent pesticides, it is still toxic, and should be placed where pets and children can't get to it.

- Follow ants back to where they're getting in and block their entry with chili powder.

- Instead of using toxic products, try homemade substitutes made with common, less threatening ingredients. (See "Simple Substitutes for Toxic Products," page 46.)

- Use old-fashioned fly paper instead of chemical pesticides.

- If you have a flea problem indoors, try a flea trap before resorting to pesticides. Available from "Gardens Alive!," the trap attracts fleas with a low-wattage light to a sticky surface which traps them permanently.

- Make your own flea trap by suspending a light bulb 10 inches over a pan of soapy water. The fleas leap for the light only to fall into the water and drown.

A Wildlife Wonderland

RESOURCES

BUGS, SLUGS & OTHER THUGS
Controlling Garden Pests Organically
by Rhonda Massingham Hart

THE NATURAL HOUSE BOOK
by David Pearson

THE TOXICS PROJECT
Florida Defenders of the Environment
2606 N. W. 6th St.
Gainesville, FL 32609
(904) 372-6965

Provides technical assistance to grassroots groups and individuals.

THE BIO-INTEGRAL RESOURCE CENTER
P. O. Box 7414
Berkeley, CA 94707

Send a self-addressed, stamped envelope for a list of their publications to answer questions you have about controlling pests without toxic products.

THE WASHINGTON TOXICS COALITION
4516 University Way, NE
Seattle, WA 98105

Send a self-addressed, stamped envelope for their list of fact sheets about least-toxic alternatives to household products.

EARTH VISION
2721 Forsyth Rd., Suite 367
Winter Park, FL 32792
800-327-8423

Sells environmentally responsible garden products.

SEVENTH GENERATION
Colchester, VT 05446-1672
800-456-1177

Sells environmentally responsible household products.

THE BENEFICIAL INSECT CO.
244 Forrest St.
Fort Mill, SC 29715
(803) 547-2301

Sells beneficial insects for the garden or farm.

GARDENS ALIVE!
P. O. Box 149
Sunman, IN 47041

Offers environmentally responsible products for the garden, including beneficial insects, insecticides made from plants and minerals, flea controls, and insect traps.

SIMPLE SUBSTITUTES FOR TOXIC PRODUCTS

The average household contains common substances, relatively free of toxic effects, that can do the work usually done by more hazardous products. Give these a try.

- To OPEN CLOGGED DRAINS, first pour in one half cup baking soda, follow that with one half cup salt, and then add one quarter cup vinegar. Cover the drain until the fizzing stops, and finish the job by pouring in boiling water. If that sounds like too much to remember, use a plunger or snake.

- CLEAN WINDOWS and GLASS SHOWER DOORS with a mixture of two tablespoons white vinegar and one quart warm water. A little extra vinegar may be helpful on the shower door if your water has left lime deposits — a common problem in Florida.

- If grease is a problem on your MIRRORS and WINDOWS try this solution: mix three tablespoons ammonia, one table-spoon white vinegar, and one cup water in a spray bottle.

- To REMOVE RUST or LIME DEPOSITS on the outside of win-dows, use a mixture of one-quarter-cup cornstarch, one-half-cup ammonia, and one cup vinegar.

- Use lemon juice to POLISH BRASS; lemon juice and salt to POLISH COPPER.

- To CLEAN YOUR OVEN, pour one quarter cup ammonia in a non-aluminum pan along with water to cover the bottom. Place the pan in your oven, still warm from baking (but off) and leave it overnight. Scrub the softened residue off with baking soda and water.

- REMOVE MILDEW with vinegar.

- A good GENERAL PURPOSE CLEANER can be made by mixing two tablespoons of ammonia with two tablespoons of liquid detergent in one quart of warm water.

- Use leftover soap slivers to make your own SOFT SOAP for handwashing. Dissolve one cup of shaved soap in one quart of boiling water. When it's completely melted or dissolved, pour it into a wide mouth jar. It will jell as it cools.

- MAKE DUST CLOTHS that gather rather than scatter dust. Put cloths in a jar or tin container containing a few drops of oil or wax. Leave them overnight in the covered container. By morn-ing your cloths will have absorbed just enough oil or wax to dust and polish at the same time.

- GET RID OF STATIC ELECTRICITY with a spray of fabric softener and water.

- For INSECT CONTROL IN THE GARDEN, the U.S. Department of Agriculture recommends making a concentrate of one tablespoon of dish washing liquid added to a cup of vegetable oil (corn, soybean, peanut, safflower or sunflower). To use, mix two-and-a-half-teaspoons of this mixture to one cup of water in a spray bottle. This has proven effective against aphids, whiteflies, and spider mites.

- TO DETER PLANT PESTS on house plants or in the garden, try a spray made by mixing a tablespoon of red pepper sauce with one-quarter cup-dishwashing liquid and a quart of water.

A Wildlife
Wonderland

RESOURCES

ALTERNATIVES THAT ARE RELATIVELY
   FREE OF TOXIC EFFECTS
   Institute of Food and Agricultural Sciences
   University of Florida
   Gainesville, FL 32608
   Available from your county extension agent.

See "Protect People, Pets and Wildlife from Poisons," page 42 and "Keep Contaminants Out of Our Drinking Water," page 88.

DON'T KILL WILDLIFE WITH KINDNESS

The Endangered Species Act makes it a crime to do anything that alters the behavior of endangered species. Even so, people with the best intentions continue to harm wildlife, mistakenly believing they're being helpful. Artificial feeding is probably the most frequent violation. "Rescuing" healthy baby birds or animals also is common.

WHAT NOT TO DO

- Don't feed wild alligators. It's not only dangerous and illegal, it's cruel. A gator that associates people with food loses its fear of them and may become a threat. When that happens it may have to be destroyed, or at the very least, captured and relocated.

- Don't feed gulls and other shorebirds. They are more likely to get entangled in fishing lines when they learn to come to people for food.

- Don't rely on artificial feeding stations to attract birds. Clean, well-kept birdfeeders are fine, but natural food sources also offer shelter and cover without altering behavior. Find out what plants bear seeds, berries or fruit preferred by birds in nearby natural habitats, and cultivate these natives in your yard or garden. (Read "Landscape for Wildlife" on page 36.)

- Don't "adopt" baby birds. Most apparently orphaned birds are under the watchful eyes of a parent. If it appears that the bird has fallen from the nest and isn't trying to fly, try to find the nest and replace the bird. (It's not true that the parent will reject a baby handled by humans.) If you find an injured bird, call a veterinarian or wildlife rehabilitator. You may need to care for it temporarily, but remember that our state and federal laws prohibit the permanent possession of wild birds.

RESOURCES

FLORIDA WILDLIFE REHABILITATOR'S ASSO.
 Miami Museum of Science
 3280 S. Miami Ave.
 Miami, FL 33129-9989
 (305) 854-4242
 Contact to find a wildlife rehabilitator near you.

(407) I C EAGLE, that is (407) 423-2453.

Call to report an injured or orphaned bird of prey.

SUNCOAST SEABIRD SANCTUARY INC.
18328 Gulf Blvd.
Indian Shores, FL 33535
(813) 391-6211

Send a self-addressed, stamped #10 envelope for two pamphlets, *Help for Hooked Birds* and *Care and Feeding of Orphan Song and Garden Birds.*

A Wildlife Wonderland

PUT THE BRAKES ON ROAD KILLS

- A recent study found that 13,000 snakes weighing a total of 1.3 tons were killed on a 2.9-mile stretch of U.S. 441 that crosses Payne's Prairie State Preserve near Gainesville.

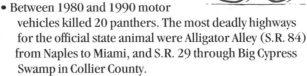

- Between 1980 and 1990 motor vehicles killed 20 panthers. The most deadly highways for the official state animal were Alligator Alley (S.R. 84) from Naples to Miami, and S.R. 29 through Big Cypress Swamp in Collier County.

- Rural roads in Polk and Osceola Counties, where the state's largest populations of eagles are found, are the most deadly for the national bird. Collisions with cars and trucks claimed 56 from 1980 to 1990.

- More than 400 key deer were killed crossing U.S. 1 on Big Pine Key between 1980 and 1990.

- Crocodiles may seem like unlikely traffic victims, but they are vulnerable during mating season when they regularly cross U.S. 1 in the upper Florida Keys.

- S.R. 40 in Marion and Lake Counties, S.R. 29 through Big Cypress Swamp, and U.S. 19 from its intersection with S.R. 50 north to the Hernando-Citrus County line are frequently crossed by black bears. At least 234 didn't make it between 1980 and 1990.

Highway traffic, and roads themselves where they pass through valuable habitats, are major threats to some of Florida's best-known imperilled animals. Panthers, eagles, key deer, black bears, and even crocodiles are frequent victims of collisions with motor vehicles. Cars and trucks take a heavy toll on snakes, wood storks, scrub jays, and gopher tortoises as well. These and other animals are also threatened when roads separate them from critical parts of their former environment, or others of their kind.

For example, gopher frogs living in upland sandhills may be unable to reach the wetlands necessary for breeding, when the two ecosystems are separated by a highway. In some cases, smaller habitats lead to genetic problems caused by inbreeding. We must pay careful attention not only to how we drive, but to where we build roads, if we hope to put a stop to road kills.

SUCCESS STORIES

- Concrete tunnels beneath roads allow wildlife to cross from one side to the other safely. At $500,000 each, they are not frequently built. The Florida Department of Transportation and the Fresh Water Fish and Game Commission are developing a plan to identify areas most in need of tunnels for wildlife.

- Underpasses are being constructed for panthers on Alligator Alley and S.R. 29 in Collier County's Big Cypress Swamp.

A Wildlife Wonderland

WHAT TO DO

- Reduce your speed and be alert while driving in areas where wildlife may cross. Be especially careful in areas with signs warning of wildlife crossing.

- Write to the Department of Transportation and ask that speed limits be reduced and warning signs be placed on roads frequently crossed by wildlife.

- Speak out against proposed roads through sensitive habitats.

RESOURCE

FLORIDA DEPARTMENT OF TRANSPORTATION
    605 Suwannee Street, MS 54
    Tallahassee, FL 32399-0450
    (904) 488-3111

    Write asking for the reduction of speed limits and the placement of warning signs on roads frequently crossed by wildlife.

REDUCE LITTER'S TOLL ON WILDLIFE

- 260 miles of fishing line were found on Florida beaches during the 1990 one-day beach cleanup campaign sponsored by the Center for Marine Conservation.

- Balloons filled with helium and released at celebrations find their way into the ocean where they are mistaken for jellyfish or other food and eaten by fish, birds, sea turtles, and marine mammals.

  - The Florida Department of Transportation spends $2.8 million each year to clean up the 10 percent of Florida roads for which it's responsible.

  - The annual volume of highway trash in Florida is estimated to be enough to fill the Orange Bowl four times.

Litter is a lot worse than unsightly — it can be deadly to wildlife. Carelessly tossed plastic trash and abandoned fishing lines and nets are among the worst culprits along Florida's coastlines.

Because plastic floats on the surface, it is often mistaken for natural foods and eaten by fish, turtles, birds and other marine animals. Floating plastic bags look like jellyfish, plastic pellets like eggs, and other shapes like crabs and various prey species. If the animal doesn't choke to death, it may starve when plastic becomes lodged in its throat or stomach.

Animals entangled in fishing lines, abandoned nets or six pack rings may be unable to breathe, eat, or escape from predators. Even if the animal remains mobile, as it grows the plastic cuts into its flesh and may cause a slow, painful death.

WHAT TO DO

- In Florida it's against the law to release a bunch of 10 or more helium-filled balloons. Go a step further by not releasing even one. If a school or organization in your community is planning a balloon release for a special event, explain why you think it's wrong.

- Never leave tangled fishing line behind.

- Take an empty bag or box with you when boating or picnicking on the beach. Make a commitment to take your own trash, and any other litter you find, to an appropriate trash receptacle.

- Find out when the Center for Marine Conservation's annual beach cleanup is scheduled in your area and participate (or organize your own!). The telephone number for CMC's Florida office is (813) 895-2188.

- If you live near a coast, bay, lake, river, stream, or estuary, Adopt-A-Shore! Groups or individuals participating in this program are expected to conduct three cleanups per year for at least two years. For information on how you can take part, call Keep Florida Beautiful at 800-828-9338.

- Participate in the Adopt-A-Highway Program sponsored by the Florida Department of Transportation. Each group or individual commits to removing litter from an adopted two-mile section at least four times annually, for two years, and agrees to attend two safety meetings a year. Participants may also take part in the wildflower program by planting seeds purchased from DOT.

- If you see anyone breaking the law by dumping plastic trash into the ocean, call the Florida Office of the Center for Marine Conservation.

- Don't flush plastic down the toilet.

RESOURCES

THE CENTER FOR MARINE CONSERVATION
Florida State Office
One Beach Drive, S.E., #304
St. Petersburg, FL 33701
(813) 895-2188

Call for information about its annual beach cleanup or to report the dumping of plastic trash in the ocean.

KEEP FLORIDA BEAUTIFUL,INC.
402 West College Ave.
Tallahassee, FL 32301
800-828-9338

This organization is "dedicated to improving Florida's environment in the areas of recycling, litter prevention, pollution prevention/waste minimization, composting, and the development of markets for recycled materials and products." Write or call for information on the Adopt-A-Shore Program.

FLORIDA DEPARTMENT OF TRANSPORTATION
605 Suwannee St., M554
Tallahassee, FL 32399-0450
800-BAN-LITER or (904) 488-3111

Call for information on the Adopt-A-Highway program, or write to the attention of Steve Liner, Program Manager. Write or call for information on the Adopt-A-Highway Program.

KEEP EXOTIC ANIMALS FROM MAKING FLORIDA HOME

- Escaped or released non-native pets and domestic animals making Florida home include:

  > 16 different parrots and parakeets
  > 3 species of mynas
  > exotic fish species
  > armadillos
  > giant poisonous toads
  > iguanas
  > pythons
  > boa constrictors
  > pigs
  > domestic cats

- The Southeast Asian walking catfish escaped from a fish farm in Broward County in 1966. Able to scramble long distances on its pectoral fins and breathe air, it adapted easily to droughts that killed off other fish. Walking catfish now live in 22 Florida counties and have become the dominant fish in portions of the Big Cypress National Preserve.

- House cats make good pets, but when allowed unlimited freedom outside, they can decimate wildlife populations of small critters. According to University of Wisconsin researchers, the state's estimated one million free-ranging house cats may be killing hundreds of millions of birds and small mammals annually. A recent British study found that 70 million native animals, including 35 million song birds, are killed by cats in Britain each year.

- The Okaloosa darter, a fish unique to Florida, was thriving until the introduction of the brown darter, a competitive species. Now it appears the native will be replaced by the exotic species.

Florida has more established nonnative animals than any other state except Hawaii. South Florida's subtropical climate, and numerous tropical plant species used in landscaping around homes and parks, provide a niche for animals from other countries. Some have been introduced deliberately, others by accident. The pet trade continues to serve as a source of exotic animals, some of which escape or are released to live out their lives, and sometimes form successful breeding populations.

Some introduced animals don't presently cause severe problems, but others, such as fruit-eating birds, could become serious pests. Even those that seem like good citizens now, may eventually replace native species integral to existing native systems.

LEARNING FROM OTHERS' MISTAKES

A Wildlife Wonderland

- British colonists triggered the ecological collapse of Africa's largest freshwater lake when they introduced the Nile perch, a fast-growing carnivore, into Lake Victoria, in the 1960s. These fish decimated native species that once kept the lake clear of algae. The algae grew wild, then sank and decayed. Unchecked by native species, the snail population exploded and helped spread parasitic disease among the residents of the lakeshore.

- In Hawaii, native birds face predation from non-native species such as rats, mongoose, and wild cats. They are also threatened by disease and competition from introduced birds such as the Japanese white-eye. Several species are nearing extinction because imports such as the pig destroy the birds' habitat.

- Exotic species endanger habitats as well as individual species in Hawaii. On the island of Maui, forests overrun with wild livestock lose up to 10 centimeters of topsoil annually. The resulting silt runoff has already devastated reefs offshore.

- Mongoose introduced to control sugarcane-eating rats in the Caribbean have become a plague.

- On Guam the exotic brown tree snake has wiped out most of the native bird species.

- Common carp from Asia stocked in U.S. ponds by the government now compete with native waterfowl for aquatic plants.

- Introduced species such as rabbits and sheep contributed to the local extinction of many hairy-nosed wombat populations in Australia.

WHAT TO DO

- Have pets neutered.

- If you must give up a pet, find it a good home. Besides being unethical and cruel to release nonnative animals into the wild, it's illegal. When released pets manage to survive, they do so by preying on native animals or out competing them for food and other resources.

- Put a bell on your cat's collar, and a leash on your dog. Better yet, keep them confined during nesting season. Ground- and beach-nesting birds are especially susceptible to predation and other disturbances April through July.

- Insist on careful study and safeguards before supporting the use of exotics to control other exotic pests.

RESOURCES

THE BIRDER'S MISCELLANY
  by Scott Weidensaul

FLORIDA'S BIRDS
  Herbert W. Kale II & David S. Maehr

"INVADERS OF THE LAST ARK"
  by Nini Bloch
  Earthwatch, November 1991

"DOWN UNDER"
  by Pamela Parker
  Zoo Life, Fall 1991

HELP ROOT OUT
THE GREEN ALIENS

- The spread of melaleuca alone claims an average of 2,650 acres of wetlands a year, more than twice the 863 acres lost to development annually.

- Hydrilla causes the loss of 1,500 acres of aquatic habitat a year, double that permitted by the Department of Environmental Regulation for development.

- $100,000,000 is being spent to reduce phosphorous in discharges to the Everglades, yet this "protected" ecosystem will become a forest comprised of melaleuca and Brazilian pepper within the next 20 years if we're unable to control the woody exotics.

- Sandhill habitat is being purchased to protect the gopher tortoise, fox squirrel and other endemic species, yet these efforts will be futile if we don't stop the growth of exotic cogon grass which could cover these areas in the near future.

A Wildlife Wonderland

Some experts rank biological pollution—the spread of exotic species—as the number one environmental problem facing Florida.

Our highly urbanized state is home to a great number of exotic plants. In fact, an estimated 27 percent of Florida's named plants are non-native. That might not be much of a problem if these aliens could be confined to the landscaped areas, but our warm, wet climate provides ideal conditions for many of these introduced plants to escape from cultivation and become naturalized.

Newly introduced plants frequently have a competitive advantage over native plants, because the newcomers left their natural enemies behind, and new enemies have yet to evolve. Serious ecological changes have resulted from the intrusion of Australian pine, Brazilian pepper, Asiatic colubrina, melaleuca, and other exotics. Native Florida plant communities have already been replaced in some areas by these aggressive introduced plants that have no positive role in the food chain. In many parts of South Florida, there is not a single native plant as far as the eye can see.

THE INVASIONS

- Melaleuca appeared harmless in 1906 when it was brought to Florida from Australia as a potential lumber tree, and to help dry up the Everglades, considered a wasteland at the time. Since then it has spread so rapidly that it threatens to replace and eliminate many of Florida's native plant communities and the animals that live in them. It already infests the Big Cypress

National Preserve and the Loxahatchee National Wildlife Refuge. It has been predicted that by the year 2000, melaleuca will dominate half of South Florida.

- Asiatic colubrina, first introduced into the Caribbean Islands from Asia, escaped from cultivation to launch its attack on Florida by sea. Its floating seeds reach our uplands during spring and storm tides. Once established it grows over native vegetation and often effectively shades out our native plants.

- Australian pine was brought to Florida in the late 1800s to be used for shade, lumber and windbreaks. It grows fast, forming dense stands that crowd out native vegetation.

- We welcomed Brazilian pepper when it was introduced to Florida in the late 1800s for ornamental purposes. It is possible that this alien produces a chemical in its leaves that suppresses the growth of competing plants. This aggressive tree has taken over thousands of acres of wetlands, hammocks, pinelands and pastures.

- Following the depression, the newly created Soil Conservation Service provided millions of kudzu plants and the labor required to plant them. Southern farmers were encouraged to use it on their land vulnerable to erosion. Unfortunately, that widespread introduction, and the absence of natural diseases or insect enemies allowed kudzu to spread ferociously, smothering mature forests by blocking sunlight with its dense foliage.

- It is widely believed that the aquatic plant hydrilla was introduced to Florida waters when tropical-fish aquariums were emptied into waterways. In some cases hydrilla and other imported species may have been planted in rivers by aquarium suppliers. Another theory pinpoints the origin of hydrilla to a backyard goldfish pond in Crescent City, from which it escaped to the St. Johns River in the 1880s. It spreads when pieces clinging to boat propellers and fishing tackle hitchhike to previously uneffected areas. With a growth rate of more than an inch a day in warm weather, this alien has choked thousands of acres of lakes, rivers and canals so completely that few native plants survive.

COUNTER ASSAULT

- Florida may get the nation's first wildlife refuge for the protection of rare native plants. The area proposed is a strip of desert like scrub running down the center of the state, home to some 40 plant species found nowhere else in the world.

- Big Cypress National Preserve benefited from a U.S. Department of the Interior project that removed a heavy growth of Brazilian peppers along a 28-mile stretch.

- The Florida Chapter of The Nature Conservancy frequently invites volunteers to participate in scheduled workdays to help remove exotic plants from preserves throughout the state.

A Wildlife Wonderland

WHAT TO DO

- Use native plants in your yard.

- If you are a water gardener, don't introduce exotic plants such as hydrilla into your pond. They may be carried into natural waters by water birds.

- If you observe an alien plant species escaping from cultivation into natural areas, write the Exotic Pest Plant Council.

- If exotic plants are taking over your yard, try digging them up root and all. In the case of trees too large to uproot, you may have to use a herbicide as a last resort. If so, use a paintbrush to apply it to a cut wood surface rather than spraying.

- Don't use fertilizers near lakes or rivers. Exotic water plants thrive where well fed on lawn fertilizers that wash into waterways.

RESOURCES

EXOTIC PEST PLANT COUNCIL
c/o Florida Department of Natural Resources
2051 East Dirac Drive
Tallahassee, FL 32310
Write to report exotic plants growing in otherwise natural areas.

FLORIDA NATIVE PLANT SOCIETY
P.O. Box 680008
Orlando, FL 32868
(407) 299-1472

This membership organization is concerned with the preservation, conservation, and restoration of native plants and native plant communities in Florida. Its newsletter/magazine *Palmetto* is published quarterly.

THE NATURE CONSERVANCY, FLORIDA CHAPTER
2699 Lee Road, Suite 500
Winter Park, FL 32789
(407) 628-5887

Call to find out how you can participate in the removal of exotic plants from preserves.

ENDANGERED FLORIDIANS

"The law locks up both man and woman
Who steals the goose from
off the common, but
lets the greater felon loose,
Who steals the common from the goose."
                    —*Anon.*

Within the last 50, years development has claimed 21% of Florida's forest land and 56% of its herbaceous wetlands. Over 40% of the 668 vertebrate species that make Florida home are declining. More than 140 fish, amphibians, reptiles, birds and mammals are classified as endangered, threatened or of special concern, and the number is growing.

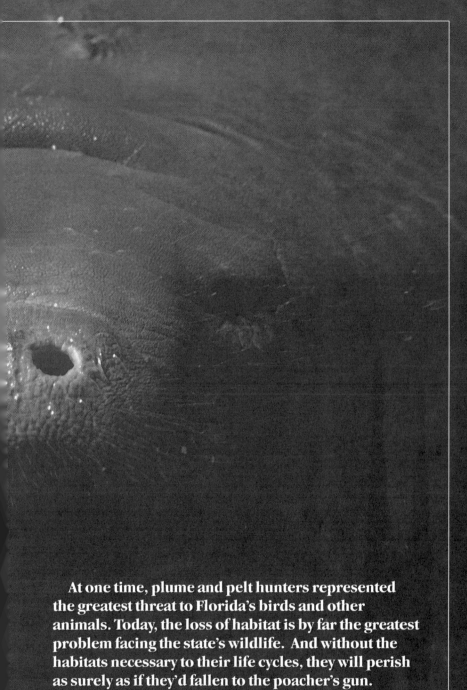

At one time, plume and pelt hunters represented the greatest threat to Florida's birds and other animals. Today, the loss of habitat is by far the greatest problem facing the state's wildlife. And without the habitats necessary to their life cycles, they will perish as surely as if they'd fallen to the poacher's gun.

In this chapter, you will learn ways you can get involved in protecting Florida's endangered species.

FLORIDA PANTHER
(FELIS CONCOLOR CORYI)

- Road kills account for the greatest number of Florida panther deaths. Twenty lost their lives on highways between 1980 and 1990.

- Eight panthers are known to have died in 1990. That's double the number of deaths in 1989.

- Poaching is the second leading cause of panther mortality, in spite of the fact it is a felony under The Florida Panther Act and The Endangered Species Act.

- Mercury poisoning has killed at least one Florida panther and is suspected as the cause of death for two others. While it's unknown what is putting dangerous levels of mercury in South Florida's wetlands, it's apparent that it is working its way up the food chain from fish to raccoons and then to panthers.

The Florida panther, our official state animal, is one of the nation's most critically endangered species. With an estimated population of only 30 to 50 some experts predict that it could become extinct in 25 to 40 years.

At one time the Florida panther ranged from eastern Texas or western Louisiana through Arkansas, Louisiana, Mississippi, Alabama, Georgia, Florida, and parts of Tennessee and South Carolina. Hunting and trapping, often for bounties, were responsible for early population reductions. Later, habitat destruction accelerated the panther's decline. Today the only remaining population is found in and around the Big Cypress/Everglades region of south Florida.

The panther's low numbers and concentration in one area place it in peril of extinction through a disease outbreak or inbreeding. Recently begun captive-breeding efforts may provide immediate insurance against extinction, and the long-term preservation of the remaining panther gene pool.

ENCOURAGING NEWS

- The federal government has established the new Florida Panther National Wildlife Refuge and has expanded national parklands in southern Florida.

WHAT TO DO

- Write your state legislators urging them to provide tax incentives for private landowners to give wildlife corridor easements to link panther ranges.

- Buy a Florida panther license plate for your car. Half of your $27 contribution will go to the Florida Game and Fresh Water Fish Commission's Panther Rescue and Management Fund. The remainder is split evenly between the Save Our State Environmental Education Trust Fund, and the Florida Communities Trust Fund.

- See "Help Keep Wildlands for Wildlife" at the end of this chapter.

Endangered Floridians

RESOURCES

FLORIDA POWER & LIGHT COMPANY
P. O. Box 078768
West Palm Beach, FL 33407-0768
800-552-8440

Write for their free booklet, *The Florida Panther.*

GAME AND FRESH WATER FISH COMMISSION
Office of Information Services
620 S. Meridian St.
Tallahassee, FL 32399-1600.
(904) 488-6661

Write for information on the captive breeding program.

WEST INDIAN MANATEE
(TRICHECHUS MANATUS LATIROSTRIS)

- Collisions with boats are so common that virtually all manatees living in Florida waters bear scars from propeller blades.

- About 100 new boats are registered in Florida daily. In addition to the more than 718,000 boats already registered in Florida, 300,000 to 400,000 more boats that are registered out of state operate in Florida waterways.

- More than 80 percent of our state's new residents settle in Florida's coastal zones. This unparalleled growth means ever-increasing coastal development, dredge-and-fill operations, and stormwater runoff — all detrimental to manatee habitat. "Mermaids" to some, "sea cows" to others, these gentle, slow-moving mammals have inhabited Florida waters for more than 45 million years. Manatees evolved with no natural enemies, and were plentiful in Florida waters until their numbers were depleted through hunting.

Today human activities remain the greatest threat to their survival, accounting for more than half of manatee deaths with identifiable causes. Boat and barge collisions, crushing or drowning in flood gates, poaching, ingestion of fish hooks and monofilament line, and entanglement in crab trap lines are constant threats.

But ultimately, we must do more than slow boats, remove unnecessary water control devices, enforce laws against hunting manatees, and pick up litter. To save manatees, we must protect their habitat.

WHAT TO DO

- Stay out of designated manatee sanctuaries.

- Never drag your boat propeller through seagrass beds. If you do you'll destroy critical habitat for manatees and many other marine species.

- Observe posted speed-zone signs when boating in areas where manatees might be present.

- Stay at least 50 feet away from manatees when you're operating a power boat. Cut the engine and paddle slowly and quietly into position to watch them.

- Never harass a manatee or participate in any activity that may change its natural behavior. Follow these guidelines when swimming near manatees:

    Enter the water slowly and quietly.

    Look, but don't touch unless the manatee approaches and touches you first. Only then is it permissible to touch gently with an open hand.

    Never chase a manatee. If it avoids you, you should avoid it. Your pursuit could drive it into marginal habitat and threaten its survival.

    Don't come between an individual manatee and its group.

    It's especially important not to separate a cow and her calf.

    Don't try to feed, grab, hold or ride a manatee.

- Always take the time to retrieve tangled or broken fishing line. Monofilament line and hooks hidden in the aquatic plants manatees eat may cause painful injuries or even death.

- Pick up any litter you find along waterways. Better yet join or organize a group cleanup campaign to rid a local river or stream of potentially dangerous trash.

- Call 800-342-1821 to report dead, injured or tagged manatees, or to report someone who is breaking the rules for protecting manatees. This "Manatee Hotline" is operated by the Florida Department of Natural Resources.

- Buy a manatee license plate for your car. The $17 added to your license fee will be split between the Save Our State Environmental Education Trust Fund and the Save the Manatee Trust Fund housed in the Department of Natural Resources.

- Join the nonprofit Save the Manatee Club and participate in their Adopt-a-Manatee Program.

- Help preserve habitat. See "Help Keep Wildlands for Wildlife" at the end of this chapter.

Endangered Floridians

RESOURCES

FLORIDA POWER & LIGHT COMPANY
Corporate Communications
P.O. Box 029100
Miami, FL 33102

Ask for *Boaters' Guide to Manatees: The Gentle Giants.*

FLORIDA POWER & LIGHT COMPANY
P. O. Box 078768
West Palm Beach, FL 33407-0768
800-552-8440

Write or call for the free booklet The West Indian Manatee in Florida. FP&L also offers a 16-mm film, Silent Sirens: Manatees in Peril to schools.

SAVE THE MANATEE CLUB
500 N. Maitland Ave.
Maitland, FL 32751
(407) 539-0990

Call or write to adopt a manatee or for membership information.

FLORIDA DEPARTMENT OF NATURAL RESOURCES
Marine Research Institute
2100 8th Ave, S.E.
St. Petersburg, FL 33701
(813) 896-8626

Operates a "Manatee Hotline"—800-342-1821.

CRYSTAL RIVER NATIONAL WILDLIFE REFUGE
U.S. Fish and Wildlife Service
(904) 563-2088

Call for information on their "Manatee Watch Program," a volunteer effort to provide information to boaters and divers on how to behave around manatees.

SEA TURTLES

*"Mankind can help protect these gentle creatures and
their habitats or witness their departure from the planet."*
— *Florida Senator Robert Graham*

- Four out of five of Florida's sea turtles are classified as endangered. The loggerhead is classified as threatened, but even its population is declining.

- The number of mature, female Kemp's ridley turtles has dropped more than 99 percent in the last forty years.

- Over two dozen endangered green sea turtles washed ashore on Florida's east coast during January, 1991. Many had net marks or cuts from their fatal encounters with nets.

- Valuable sea turtle nesting habitat is being lost to coastal construction and excavation.

- Many sea turtles die each year when they eat plastic mistaken for food.

Endangered Floridians

Sea turtles have been around since the days of the dinosaurs. Now commercial fishing, ocean pollution and coastal development threaten to drive them into extinction.

Florida is host to five species of endangered sea turtles — the leatherback (Dermochelys coriacea), Atlantic green (Chelonia mydas mydas), Atlantic loggerhead (Caretta caretta caretta), Kemp's ridley (Lepidochelys kempi) and Atlantic hawksbill (Eretmochelys imbricata imbricata). More threatened loggerhead turtles nest on a 20-mile stretch of beach on the east coast of Florida, than anywhere else in the western hemisphere. This area has been designated the Archie Carr National Wildlife Refuge by the U. S. Fish and Wildlife Service. The second largest loggerhead rookery in the world, this is a habitat of global importance. Not only do one-fourth of the world's loggerhead nests occur here, nearly half the endangered green sea turtles in the United States nest here as well.

NEW HOPE

- Florida's Marine Turtle Protection Act passed in 1991 strengthens sea turtle protective measures and clarifies the urgency of sea turtle considerations in coastal construction and excavation permitting.

• According to the National Academy of Sciences, shrimp trawlers are responsible for the greatest number of sea turtle deaths. A state law, intended to save sea turtles, requires shrimpers to use Turtle Excluder Devices (TEDs) in all Florida offshore waters. TEDs consist of large-mesh webbing or metal grids inserted into the shrimp nets. As the funnel-shaped nets are pulled along the sea-bed, shrimp and other small marine animals slip through the TED and into the narrow collection end of the net. Sea turtles and other fish and animals too large to pass through are deflected out an escape hatch.

WHAT TO DO

• Be careful not to disturb a nesting turtle by making noise or shining lights. When frightened, a sea turtle may return to the sea without laying her eggs.

• Never touch or attempt to ride a sea turtle.

• When you visit the beach, watch out for hatchlings. They may be heading in the wrong direction if they're confused by lights from the road or development.

• Make a commitment to leave nothing but footprints on the beach. Pick up your own trash and any left by others.

• If you find an injured or dead turtle, or witness abuse, call the Florida Marine Patrol in your area, or call the Florida Department of Natural Resources toll-free Resource Alert Line, 800-342-1821 to report it.

• If you find a turtle with a metal tag, report the tag number to any Marine Patrol officer, or call the Resource Alert Line (800-342-1821).

• Attend hearings on local ordinances on protecting sea turtles.

• Learn what lands are being considered for state purchase by the Florida Conservation and Recreational Lands (CARL) program. Write or call members of the CARL Selection Committee urging them to rank sea turtle nesting beaches high on their list of land acquisition priorities.

• Adopt a turtle through the Caribbean Conservation Corporation. Call 800-678-7853 for information.

RESOURCES

CARIBBEAN CONSERVATION CORPORATION
P. O. Box 2866
Gainesville, FL 32602
800-678-7853

Call for information about adopting a turtle.

THE CENTER FOR MARINE CONSERVATION
Florida State Office
One Beach Drive, S.E., #304
St. Petersburg, FL 33701
(813) 895-2188

DEPARTMENT OF NATURAL RESOURCES
3900 Commonwealth Blvd.
Tallahassee, FL 32399-3000
800-342-1821 (Resource Alert Line)

Call to report sightings of tagged turtles.

FLORIDA POWER & LIGHT COMPANY
Corporate Communications
P. O. Box 029100
Miami, FL 33102-9100

Write for their free booklet, *Florida's Sea Turtles*.

NATIONAL AUDUBON SOCIETY
550 South Bay Ave.
Islip, NY 11751

Ask for *Conserving Marine Resources*.

CONSERVATION AND RECREATIONAL LANDS PROGRAM
3900 Commonwealth Blvd.
Mail Station #5
Tallahassee, FL 32399-3000

Write urging the acquisition of sea turtle nesting beaches.

Endangered
Floridians

KEY DEER
(ODOCOILEUS VIRGINIANUS CALVIUM)

- More Key deer are killed by traffic each year than are born.

- Well-meaning but thoughtless people who feed the deer also contribute to their decline by teaching them to accept unnatural foods, and not to fear humans. Several have died with plastic bags in their digestive tracts.

- Human cruelty has also taken its toll. In 1991 two men were convicted of clubbing a doe to death.

This native of the Florida Keys is a miniature relative of the Virginia white-tailed deer unique to our state. Cut off from the mainland, it adapted to life among palmettos and mangroves, evolving into an animal different from its larger relatives. Sometimes referred to as "toy deer," a full-grown key deer stands a mere 26 to 32 inches at the shoulder. Adult females weigh between 35 and 65 pounds, and males weigh up to 90 pounds.

Key deer inhabit islands in the lower Keys between Sugarloaf Key and Johnson Key, and prefer pine and hardwood forests. They commonly browse on tender leaves and shoots of trees, shrubs, and herbs in young forest stands and in clearings for right-of-ways.

In the '50s there were only 25 to 50 of these deer left due to poaching. From that low the population built up to an estimated 450 before loss of habitat reduced their numbers to the current estimate of 250.

SUCCESS STORIES

Endangered
Floridians

- The 7,000-acre Key Deer National Wildlife Refuge was established in 1954 to protect the deer's vanishing habitat.

- Florida has reduced the speed limit at night from 45 to 35 mph on a busy highway that cuts through the main habitat of the Key deer on Big Pine Key.

WHAT TO DO

- Drive with extreme caution on roads through key deer habitat. (See "Put the Brakes on Road Kills," page 50.)

- Resist the urge to feed them, even if they approach you.

- Encourage the preservation of key deer habitat. (See "Keep Wildlands for Wildlife," at the end of the chapter.)

RESOURCES

FLORIDA'S VANISHING WILDLIFE, Circular 485
  by Laurel Comella Hendry, Thomas M. Goodwin and Ronald F. Labisky
  Available from your county extension agent.
GUIDE TO FLORIDA'S VANISHING WILDLIFE
  by Robert Anderson

FLORIDA BLACK BEAR

(URSUS AMERICANUS FLORIDANUS)

- The bear's need for space in a state criss-crossed with roads, makes it vulnerable to highway traffic. The Florida Game and Fresh Water Fish Commission reported that 33 bears were killed on roads in 1990.

- Poaching, with bear parts sold for profit, illegally claims some bears. Limited bear hunting is still allowed in parts of the state. Osceola National Forest has a nine-day bear season in January; there's a 12-day season in Apalachicola National Forest in late November and early December; and hunting is allowed on private land in Columbia and Baker Counties in late November and early December.

- Habitat destruction through residential and commercial development is the greatest threat to the bear's survival.

The Florida black bear once roamed throughout the state, but is now found in fewer than 40 counties. Considered threatened except in Baker and Columbia Counties and in the Apalachicola National Forest, it makes its home in river swamps, scrub oak areas, cypress swamps, large hammocks, and pine flatlands. The largest land mammal resident to the state, adult males reach a length of about five feet and weigh close to 300 pounds on a diet of saw palmetto and black gum berries, insects such as yellow jackets, birds, and small mammals.

Such a large animal requires a lot of space—at least a 15-mile radius home range. A full-grown male may claim a home range of 66 square miles.

WHAT TO DO

- Reduce your speed and use extra caution when driving on roads frequently crossed by bears—S.R. 40 in Marion and Lake Counties, S.R. 29 through Big Cypress Swamp, and U.S. 19 from its intersection with S.R. 50 north to the Hernando-Citrus County line. (See "Put the Brakes on Road Kills," page 50)

- Report poaching or harassment of bears to the Florida Game and Fresh Water Fish Commission at (800) 342-8105, 24 hours a day, 365 days a year.

- Help preserve bear habitat. (See "Help Keep Wildlands for Wildlife" at the end of this chapter.)

RESOURCES

FLORIDA'S VANISHING WILDLIFE, Circular 485
  by Laurel Comella Hendry, Thomas M. Goodwin and Ronald F. Labisky

  Available from your county extension agent.

GUIDE TO FLORIDA'S VANISHING WILDLIFE
  by Robert Anderson

GAME & FRESH WATER FISH COMMISSION
  620 South Meridian St.
  Tallahassee, FL 32399-1600
  800-342-8105 (Wildlife Alert)

  Call to report poaching or harassment of bears.

Endangered Floridians

FEATHERED FLORIDIANS AT RISK

More than a dozen of Florida's birds are suffering dangerous declines in their populations. In every case the primary cause of decline is habitat degradation and destruction. At highest risk are those birds with highly specialized diets or feeding habits, and those dependent on a particular land type. These five species help illustrate some of the problems our feathered friends face.

SNAIL KITE
(ROSTRHAMUS SOCIABILIS)

The snail or Everglade kite is particularly vulnerable because its diet consists almost exclusively of a genus of freshwater mollusk called apple snail, or Pomacea. The kite's slender, sharply hooked bill is ideal for extracting the living animal from the unbroken shell.

When times are tough the kite may eat an occasional small turtle or freshwater crab, but its dependence on apple snails is so nearly total that it is confined to freshwater habitats. Rising and falling water levels affect not only the availability of food, but also stable nest sites. It is no wonder that the snail kite population literally rises and falls with water levels in the Everglades.

WOOD STORK
(MYCTERIA AMERICANA)

The only North American stork, this predominantly white bird stands three feet tall with a wingspan of five feet. Its dark gray wrinkled upper neck and head, and massive long beak make it easy to identify.

Unlike most other wading birds, wood storks don't use vision to locate their prey. As the bird walks forward, the bill is slightly open and submerged up to the breathing passages. The bill is swept from side to side until a fish is touched—triggering one of the fastest reactions known in vertebrates—and the bill snaps shut.

Though this feeding technique is quite efficient, it requires a concentrated supply of fish. Naturally fluctuating water levels provide dense fish populations in shrinking ponds during the wood stork's nesting cycle when the need for food is the greatest. Man-made water-control devices often keep water levels artificially high, forcing the wood stork to search for other food sources.

FLORIDA SCRUB JAY
(APHELOCOMA COERULESCENS COERULESCENS)

The Florida scrub jay is isolated in the dry sand pine and oak scrublands of peninsular Florida. Scrub jays are cooperative breeders, living in permanent, group-defended territories.

Groups often consist of a monogamous bonded pair aided by one to six helpers, generally the pair's offspring from previous seasons. Some territories however, are occupied by pairs with only one or two helpers or none.

Now restricted to scattered, shrinking oak scrub habitats, scrub jays face an uncertain future. All available habitat is occupied, and reproductive success elsewhere is unlikely.

RED-COCKADED WOODPECKER
(PICOIDES BOREALIS)

Mature pine forests are home to the red-cockaded woodpecker. These birds can excavate nest cavities only in older trees where red-heart fungus is common. Modern forest-management practices which dictate removal of fungus infested trees, and the harvest of pines before they are fully mature, have greatly reduced the availability of cavity sites.

If habitat is not preserved, this bird will go the way of the ivory-billed woodpecker, believed to be extinct due to the destruction of dense river swamps by timbering.

BALD EAGLE
(HALIAEETUS LEUCOCEPHALUS)

Though bald eagles were once decimated by the effects of the pesticide DDT, Florida's population of 500 eagle nests is stable today. Researchers from the University of Florida's Institute of Food and Agricultural Sciences take advantage of our healthy eagle population by collecting eggs from active nests in Florida, raising the young in captivity and releasing them in other states.

This "egg recycling" is possible because the female lays a replacement clutch when her eggs are removed. This program is critical to the survival of the species since only 120 active nests are found elsewhere in the southeast.

WHAT TO DO

• Help preserve habitat. See "Help Keep Wildlands for Wildlife" at the end of this chapter.

• Adopt an injured bird of prey through the Florida Audubon Society, and help in the care and release of hundreds of others. You will receive a photograph and biography of the bird you adopt plus a subscription for *Florida Raptor News.*

- Report injured or orphaned birds of prey by calling the Florida Audubon Society's Center for Birds of Prey at (407) 645-3826.

- If you witness harassment of bald eagles, call the Florida Audubon Society at (407)I C EAGLE (423-2453).

- Report nesting eagles to the Florida Audubon Society at (407) 423-2453.

RESOURCES

FLORIDA AUDUBON SOCIETY
    Adopt-A-Bird Program
    Dept. N 460 E. Highway 436 Suite 200
    Casselberry, FL 32707
    (407) 260-8300
    Call to adopt a bird.

THE BIRDER'S HANDBOOK
    A Field guide to the Natural History of
    North American Birds
    by Paul R. Ehrlich, David S. Dobkin, and Darryl Wheye

FLORIDA'S BIRDS
    by Herbert W. Kale,II and David S. Maehr

FLORIDA'S VANISHING WILDLIFE, (Circular 485)
    County Extension Agent

GUIDE TO FLORIDA'S VANISHING WILDLIFE
    by Robert Anderson

GAME AND FRESH WATER FISH COMMISSION
    Nongame Wildlife Education Section
    620 Meridian Street
    Tallahassee, FL 32399-1600
    (904) 488-6661

    Ask for *Living Treasures of the Florida Scrub,* a poster/information brochure.

HELP KEEP
WILDLANDS FOR WILDLIFE

Have we saved our wildlife from poachers and plume gatherers, only to lose it to degradation and loss of habitat?

The most important answer to saving threatened and endangered animals in Florida is to preserve wildlands. Each of us can make a difference.

Endangered
Floridians

WHAT TO DO

- Express your views as a concerned citizen. Ask local and state planners to include habitat conservation in their plans. (See "Lobbying," page 146.)

- If you own environmentally sensitive land, consider a conservation easement. You can retain your ownership while protecting the site for the future. (See Resources, The Nature Conservancy and The Trust for Public Lands)

- Join and contribute to one or more groups involved in conservation and land acquisition such as The Nature Conservancy and The Trust for Public Land.

- Help preserve wetlands. Follow suggestions in Chapter One, under "What To Do" for wetlands page, 77.

RESOURCES

THE NATURE CONSERVANCY
Florida Regional Office
2699 Lee Rd., Suite 500
Winter Park, FL 32789
(407) 628-5887
Call or write for membership information.

THE TRUST FOR PUBLIC LAND
1310 Thomasville Rd.
Tallahassee, FL 32303-5608
(904) 222-9280
Call or write for membership information.

It's hard to believe we don't have enough water.
Florida has more available groundwater than any
other state.

The Floridan aquifer, with an area of 82,000
square miles, lies under most of the state and
extends into southern Alabama, southeastern
Georgia and parts of South Carolina. One of the
most productive aquifers in the world, it is estimated to contain as much water as the entire
Great Lakes. And intermediate aquifers overlie
the Floridan.

According to the U.S. Geological Survey, 27 first
magnitude springs (those with average flows of at
least 100 cubic feet per second) well up from the
immense limestone bed underlying the
peninsula.

However, many wetlands have already been
destroyed by development, resulting in a decline
in the quantity and quality of groundwater. Construction of impervious surfaces, such as roads,
parking lots and buildings, with drainage systems
designed to channel floodwaters away from populated areas, reduce the proportion of rainfall
which recharges the underground aquifers.

There was a time when the land surface and
holes in it were considered safe and convenient
depositories for many wastes. Only recently have
we come to realize that natural soil processes
have a limited capacity to change contaminants
into harmless substances at all, much less before
they reach our aquifers.

QUID GOLD

Water pollution is a concern of anyone who wants safe drinking water. Groundwater is now a source of drinking water for approximately 50 percent of the nation's population. Here in Florida over 90 percent of us depend on groundwater as our primary source of drinking water, and some 20 percent drink untreated water drawn directly from private wells.

In this chapter you will learn ways you can help conserve our precious groundwater supplies.

PROTECT FLORIDA'S SPRINGS FROM DEPLETION AND POLLUTION

- The Floridan aquifer conducts tremendous quantities of water throughout most of the state. This cavernous underground river is the source of numerous springs, 27 of which are first magnitude, discharging more than 100 cubic feet of water per second. The average flow of these springs combined is 9,600 cubic feet per second, or slightly more than six billion gallons a day.

- People think the springs are evidence of infinite groundwater resources, and that they could actually be tapped for drinking water. If our well water is unfit to drink, or if our wells run dry, why not get our water from the springs? This has been seriously proposed but closer examination reveals two problems; the first having to do with quality, and the second with quantity.

- The porous nature of the state's topography which allows for the storage of water, also makes that water vulnerable to contamination. If the contamination we find in our well water is widespread, we'll find it in our spring water too. And it's far more expensive to meet treatment requirements for surface water than for groundwater.

- Man's demands on our aquifers can have devastating effects on the flow of springs. When large volumes of water are drawn from wells near a spring, water levels in the aquifer may fall below the orifice of the spring and stop its flow.

- Some springs in heavily populated south Florida show diminished flow.

- The flow of Kissengen Spring near Bartow in Polk County stopped when wells tapped the same aquifer. When withdrawals ceased for several months the flow began again, only to stop once more when the wells resumed pumping.

- In Texas, man's ever-increasing demand for water has reduced or stopped the flow of many springs, including two of the four first magnitude springs found in the state in 1947. Florida's springs are among the state's principal natural and scenic resources. It is believed that no other state or nation has as many major springs discharging as much water. We can thank our state's karst topography. The limestone and dolomite layer underlying Florida's topsoil is easily dissolved by rain water.

This thick, porous rock layer, known as the Floridian aquifer, acts as a sponge to store water—more than is stored in any other eastern state.

Springs result when water is discharged as natural leakage or overflow from an aquifer, through a natural opening in the ground. Some openings are small, resulting in a slight trickles or seepage. Others are so large that they form the headwaters of large rivers. The two general types are water-table springs with small and variable flows, and artesian springs in which water under pressure is forced to the surface.

WHAT TO DO

- Conserve water. See "Save Up to 80 Percent of Outdoor Water Use," page 82, "Reuse Water," page 84, and "Turn Off the Tap on Indoor Water Waste," page 87.

- Keep springs free of pollution. See "Keep Contaminants Out of Our Drinking Water," page 86.

Buried Liquid Gold

RESOURCES

SPRINGS OF FLORIDA
State of Florida,
Department of Natural Resources

FLORIDA'S GROUNDWATER RESOURCE:
Vast Quantity, Good Quality?
(Circular 944)
Institute of Food and Agricultural Sciences
University of Florida
Available from your county extension agent.

SAVE UP TO 80 PERCENT OF OUTDOOR WATER USE

- Florida uses more water for irrigation than the total used by all other eastern states combined.

- The average Floridian uses about 175 gallons of water a day, an extravagant amount compared to the nationwide average of 110. It seems even more wasteful when we look at the efficiencies achieved in western Europe. In Great Britain and Sweden the average use is 59 gallons, while France and West Germany hold the line at a mere 39.

- The creative use of drought-resistant plants can save up to 80 percent of the irrigation water routinely used. The key is to cluster plants with similar sunlight and water requirements.

Water conservation is becoming more and more important in Florida. In fact it's the law. Homeowners in each of the five water authority districts are under various outdoor water use restrictions. And with good cause.

With outdoor irrigation accounting for more than half of our water use in Florida, the yard is a good place to start conserving.

WHAT TO DO

WHAT TO DO

- Use drought-resistant plants. Plan your landscape taking your lifestyle into account. Keep only as much lawn as needed for recreation, since grass requires more water and maintenance than other parts of the landscape.

- Have your soil analyzed to determine if you need to add anything to improve soil moisture penetration and water holding capacity. For example, sand added to clay will improve its ability to drain properly, while organic materials added to sandy soils will aid in the retention of water.

- Dedicate a large area of your property to natural zones of native plants adapted to the wet and dry extremes of the Florida climate. These species usually thrive with little or no irrigation.

- Create a second drought-tolerant zone for native or cultivated plants that require watering only occasionally.

- Keep those plants that require frequent irrigation to a minimum, and group them together in "oasis" zones. This eliminates the need for watering the entire landscape to satisfy the needs of the thirsty plants.

- Spread mulch around plants and trees to slow evaporation. Pine needles and oak leaves work well and don't cost anything but the energy to gather and spread them.

- Install a drip-irrigation system or use a soaker hose to deliver water directly to the root zone where it's needed. This eliminates most water loss due to evaporation.

- If you must use a sprinkler, adjust it to avoid wasting water on sidewalks, driveways, streets, and gutters.

- Raise lawnmower blades to at least three inches for St. Augustine or Bahia grass. This allows roots to grow deeper, shading the root system and holding soil moisture better than a closely clipped lawn. Grass in shade should be mowed even higher.

- Use less fertilizer. It promotes growth, increasing the need for water.

RESOURCES

THE ASSOCIATION OF FLORIDA NATIVE NURSERIES
  P.O. Box 1045
  San Antonio, FL 33576
  (904) 588-3687

  To find out where to obtain Florida native plants, ask for the Plant and Service Locator. Also offers *Xeric Landscaping With Florida Native Plants* for $10 plus $2.50 postage.

THE NATIONAL XERISCAPE COUNCIL, INC.
  P.O. Box 163172
  Austin, TX 78716-3172.

  Write for more information on water conservation methods.

WATER MANAGEMENT DISTRICTS

Northwest Florida Water Management District
  Rt. 1, Box 3100
  Havana, FL 32333
  (904) 539-5999

St. Johns River Water Management District
  P. O. Box 1429
  Palatka, FL 32078-1429
  (904) 329-4500

South Florida Water Management District
  P. O. Drawer 24680
  West Palm Beach, FL 33416
  (407) 686-8800

Southwest Florida Water Management District
  2379 Broad St.
  Brooksville, FL 34609-6899
  (904) 796-7211

Suwannee River Water Management District
  Rt. 3, Box 64
  Live Oak, FL 32060
  (904) 362-1001

  Ask your water management district for a list of drought resistant plants suitable for your area.

Buried
Liquid
Gold

REUSE WATER

- Eighty percent of Florida's residents live near the coast, where water supplies are vulnerable to salt water intrusion due to overdraft. Already some communities pipe water from great distances or use desalination to make it drinkable.

- Due to heavy groundwater pumping, the decrease in the volume of fresh water in the Biscayne Aquifer serving Dade County has been so great, that at times wells must be closed due to salt water intrusion.

- In Hillsborough County, an 800-square-mile area of freshwater aquifer is filled with salt water from Tampa Bay.

- The amount of water needed to support the average American is more than eight times that required at the turn of the century. The increase can be attributed to cropland irrigation, use of synthetics needing water in their manufacture, use of appliances powered by electricity produced at water-hungry power plants, and the watering of lawns.

Florida's days as the land of inexpensive, limitless water are over. Due to a soaring population, the state is unable to provide all the water people want even in years of heavy rain.

Wastewater reuse will be one of the answers to water shortages, especially in coastal cities. Billions of gallons of sewage are dumped into the oceans daily. This practice is not only bad for marine animals and beach-goers, it's a terrible waste of water that could be treated and used again on lawns and golf courses.

The concept of water reuse requires that we stop thinking of used water as waste-water, and recognize it as an important resource that we should be "recycling"! It's really not a new idea; nature has been doing it all along.

Water evaporates from lakes and oceans becoming clouds; rain falls to form rivers and streams; some sinks into the earth and becomes groundwater: some finds its way back to sea. . . and so on.

Any time we're able to reuse water we reduce the demand on valuable ground water which is suitable for drinking. It doesn't make sense to irrigate lawns, golf courses, citrus groves and pasture land with drinking-quality water. It's especially senseless to flush our toilets with it, or use it to put out fires or control dust at construction sites.

SUCCESS STORIES

- In Altamonte Springs reclaimed water is used for irrigation of residential lawns and other green space areas, as well as in a car wash.

- More than 500 properties and a golf course are irrigated with about three million gallons of reclaimed water daily in Cocoa Beach.

- The Loxahatchee River Environmental Control District reuses more than three million gallons of water daily to irrigate seven golf courses.

- An artificial wetland has been developed in Orlando using water from its Iron Bridge treatment plant. It's designed as a recreational area for hiking, jogging, and nature observation.

- The city of Naples uses about five million gallons of reclaimed water daily to irrigate nine golf courses.

- St. Petersburg has reused water since 1977. A separate piping system supplies reused water for landscape irrigation of more than 6,000 residential lawns and for fire protection.

- Tallahassee has used reclaimed water for agricultural irrigation since 1966. Corn and soybeans are among the fodder crops grown, providing food for cattle during some parts of the year.

- In Orlando, water from homes and businesses is purged of viruses and other health hazards, and then pumped to irrigate citrus groves, reducing reliance on groundwater. Surplus water is pumped into ponds and replenishes the aquifer. This is the largest recycling project of its kind dedicated to agriculture.

Buried Liquid Gold

WHAT TO DO

- Urge your city and county officials to institute water reuse programs if they are not already in place.

- If you plan to build a new home, look into the feasibility of having a gray water system installed.

RESOURCES

REUSE OF RECLAIMED WATER
Booklet from Florida Department of Environmental Regulation

TROUBLED WATERS
A series, published the first week of March 1991 by The New York Times Company. Participating newspapers were: Sarasota *Herald-Tribune*, *The Ledger* of Lakeland, Gainesville *Sun*, Ocala *Star-Banner*, *The Daily Commercial* of Leesburg, Palatka *Daily News*, Lake City *Reporter*, *News-Sun* of Sebring and Avon Park, *The Newsleader* of Fernandina Beach and the Marco Island Eagle.

KEEP CONTAMINANTS OUT OF OUR DRINKING WATER

- Florida places last among the states in groundwater quality due to persistent pesticide contamination of groundwater, according to The Green Index. The report is published by a non-profit group and is based primarily on federal statistics.

- Florida has roughly 10 percent of the septic tanks in use in the United States.

- Urban stormwater resulting from the first half-inch of rain usually has a pollution content greater than that of raw sewage.

- Florida has more than 15,000 impoundments and drainage wells directing stormwater into the very aquifers that supply our drinking water.

- One quart of oil can contaminate 250,000 gallons of water.

- Six of Florida's 39 "Superfund" sites are landfills. All have contaminated groundwater.

- Most of Florida's 300 active and 500 inactive landfill sites are unlined, increasing the possibility that rainwater percolating through them may carry harmful chemicals to the ground water.

- Since 1983, water from 1,000 public and private wells in Florida has been found to contain levels of a soil fumigant, ethylene dibromide (EDB), above the state's allowed level. This widespread contamination has been found in 22 counties.

- A study of public water supplies in Broward, Dade and Palm Beach Counties showed that several sources supplying over a quarter-million people contained VOC (volatile organic compounds) concentrations in excess of Florida drinking-water standards.

- City wells in Pensacola, Gainesville, and Tallahassee have been taken out of service temporarily due to VOC contamination.

- The greatest frequency of gasoline contamination in groundwater has occurred in Dade, Broward, and Palm Beach counties. The most environmentally harmful and financially significant incident was the leaking of 10,000 gallons of gasoline, which contaminated the public water supply for 2,000 Belleview residents in Marion County, between October 1979 and March 1980.

Keeping our underground water supplies clean is not an easy task with the water table at or near the surface in much of the state. Clearly in Florida, serious water quality problems are posed by disposing of wastes in landfills. Septic tanks used in unsuitable places, and stormwater runoff also pose major problems.

As you might expect, most of the state's inactive waste sites, and a large number of its active hazardous-waste producing facilities are linked to the chemical and petroleum industries. Industrial sources of pollution include oil refineries, battery recyclers, chemical and pesticide manufacturers, electroplaters, wood preservers, mining, and industries that use solvents. Uncontrolled disposal sites containing hazardous wastes and contaminants are potential threats to our groundwater supplies.

Other hazardous waste sites were once municipal landfills that have become dangerous through years of gradual accumulation of discarded toxic household materials. Insect repellents, fertilizers, oven cleaners, bleach, spot removers, anti-freeze, and batteries are among the culprits.

Buried Liquid Gold

Though septic tanks work in well-drained soils, they are frequently used in waterfront developments where groundwater levels are high. Even when properly sited, they require periodic maintenance to prevent septic pollution.

Polluted stormwater runoff seeps into our groundwater through the sandy soils often favored for development. Because water in our aquifers often flows only a few feet per year, contamination may go undetected for years after the water supply has been severely degraded.

SUCCESS STORIES

- The U. S. Environmental Protection Agency calls Florida's toxic-waste inspection program the best in the Southeast.

- Florida led the nation as the first state to adopt a groundwater protection rule in 1984.

WHAT TO DO

WHAT TO DO

- If you change your own automobile oil, save the used oil in an empty plastic milk jug or bleach bottle and take it to an oil-collection center. K-Mart and Sears Automotive Service centers accept used oil.

- Use solar-powered batteries or rechargeables instead of disposable batteries. Save old disposable batteries until your county has a hazardous-waste disposal program.

- Give leftover paint to someone who can use it, or save it until it can be disposed of properly.

- Save unwanted or banned pesticides, herbicides, fungicides, solvents and pool chemicals until your community sponsors a hazardous waste disposal program.

- Use biodegradable, phosphate-free, non-toxic cleaning and laundry products.

- If you must use disposable diapers, flush the contents before tossing them in the trash. Too many reach our landfills full of raw sewage, which may slowly leak through the earth and pollute our water supplies.

- Learn all you can about points of contamination in your county, and join or organize a group to create community awareness. See "Volunteer Opportunities," chapter 8, for details on how to get started.

RESOURCES

FLORIDA'S GROUNDWATER RESOURCE:
Vast Quantity, Good Quality? (Circular 944)
Institute of Food and Agricultural Sciences
University of Florida

Available from your county extension agent.

WELLHEAD PROTECTION PROGRAMS:

Tools for Local Governments
U.S. Environmental Protection Agency

GROUND-WATER SUPPLY PROTECTION
Southwest Florida Water Management District
Brooksville, Florida

PROTECTING THE NATION'S GROUNDWATER
FROM CONTAMINATION
Summary report from he Technology Assessment Board, 98th Congress

TURN OFF THE TAP ON INDOOR WATER WASTE

- By Florida law, toilets used in new construction can use no more than 3.5 gallons of water per flush. A proposed U.S. law would limit us to 1.6 gallons. Most existing toilets require six gallons.

- Florida law limits shower fixtures installed during new construction to three gallons of water per minute. That amount will be reduced to 2.5 if a proposed federal law passes. Most existing fixtures require 4.7 gallons per minute.

- Severe problems already exist in some parts of Florida. The Southwest Florida Management District no longer issues water permits in an area encompassing southern Hillsborough County and parts of Manatee and Sarasota counties. Groundwater levels have dropped up to 60 feet since pumping began in this area, known as the "red hole".

Buried Liquid Gold

Residents in Marin County, California, are restricted to using no more than 50 gallons of water a day per person—less than a third of that used by the average Floridian. Why should we worry about what's happening on the far side of the continent? Because it could happen here.

Florida has a lot in common with California, and the similarities increase daily. Our population continues to grow at a rapid pace, straining the state's infrastructure and placing ever-greater demands on our natural resources.

LEARNING FROM OTHERS

- Town officials in Morro Bay, California, have found a way to allow growth after living with a decade-long new-building moratorium due to a shortage of fresh water. A builder may earn a permit to build one new residence by installing water-saving devices in 10 older homes. The retrofitted houses conserve as much water as the new one uses.

- The cities of Los Angeles, Santa Monica, and Santa Barbara offer rebates ($100 per head) to residents who replace old five-gallon toilets with new low-volume, 1-1/2 gallon models.

Homeowners in Santa Monica who don't install the more efficient toilets pay a $2 surcharge every month. The money collected goes into the rebate program. Old toilets are crushed and used as road material.

WHAT TO DO

• Install low-flow shower heads to reduce water use by 75 percent and still enjoy an exhilarating spray.

• Turn off the shower while you lather your hair and body, then turn it back on to rinse.

• Turn off the tap water while brushing your teeth or shaving.

• Don't flush as often.

• Reduce the amount of water you use with each flush by installing toilet dams (see Resources), fill a plastic bottle with water and place it in the tank, or add a couple of bricks.

• Run only full loads in the washer and dishwasher.

• Use the garbage can instead of the garbage disposal. Better yet, compost.

• When washing dishes by hand, use short blasts instead of letting the water run for rinsing. This may save up to 500 gallons a month. With double-sided sinks, fill one side with rinse water.

• Stop dripping faucets. A leak that fills a 16-ounce container in an hour wastes more than 1,000 gallons of water in a year. Turn off the shut-off valve until you have time to install a new washer or stem. Not only will this save water, the inconvenience of having to open the valve each time you use the tap will remind you to fix it.

• Say no thank you to unwanted water in restaurants.

RESOURCES

A GUIDE TO REDUCING HOME WATER USE (HE 3085)
Institute of Food and Agricultural Services
University of Florida, Gainesville

This circular is available from your county extension agent.

ST. JOHNS RIVER WATER MANAGEMENT DISTRICT
Office of Information and Communication
P. O. Box 1429
Palatka, FL 32077
(904) 328-8321

Ask for *Water Wise Living,* a brochure of home conservation tips, and *Home Water Use*, a family survey parents and children can use to learn more about their water-use habits.

ECOLOGICAL WATER PRODUCTS
Suite 18, Department GM
266 Main St
Medfield, MA 02052

Sells products to conserve water.

SEVENTH GENERATION
Colchester, VT 05446-1672

Sells toilet dams and low-flow shower heads among other environmental products.

REAL GOODS TRADING COMPANY
966 Mazzoni St.
Ukiah, CA 95482
800-762-9214
FAX: (707) 468-0301

Buried Liquid Gold

LAKES AND RIVERS

Florida has more lakes than any other state in the United States. In fact, the state contains more than 7,712 lakes and 1,700 rivers, measuring over 10,550 linear miles. According to national research, 99 percent of the population lives within 50 miles of a publicly owned lake. One-third of Americans live within five miles of a lake. The Sunshine State exemplifies this trend. The problem is how do we live near the lakes we love and keep them clean and attractive? In this chapter, you will learn ways you can help preserve Florida's lakes and rivers.

THE DEATH OF A LAKE

- Florida lakes have been determined to be the second most-polluted in the nation.

- The most severely polluted and fourth-largest lake in Florida is Lake Apopka. No swimming or fishing has occurred in the lake for over a generation. The lake has acquired muck as deep as 49 inches in some areas, largely due to wastewater dumping from nearby farms. The St. John's River Water Management District is working on a $19 million clean-up program to restore Lake Apopka. It is not difficult to see that protection and pro-active measures are much less costly than the costs of restoration.

Nutrients speed up the life cycle of a lake. When people add nutrients to a lake, they cause the premature death of a lake. The lake experiences increased algal growth. A reduction in oxygen and sunlight reaching the bottom caused by the algae can eliminate game fish and beneficial native water plants. Water quality is degraded and mosquitoes thrive. Eventually the lake becomes a marsh or bog. Inorganic nutrients come from products such as phosphate detergents, fertilizers, and other phosphorus products.

SUCCESS STORIES

- Florida has many lake success stories. In Tallahassee, trees have been planted around Lake Jackson in order to consume nutrients and provide wildlife habitat. State workers at Ribault River are removing harmful sediments and putting them on agricultural lands to provide fertilizer. More than 50 projects like these are taking place throughout Florida utilizing "pollution recovery trust funds."

- In 1973, New York state banned the use of phosphate detergents. Later reports showed a 50 to 60 percent reduction in phosphorus discharges at three large sewage treatment plants. Some counties in Florida have similar bans.

- Farmers are beginning to utilize innovative techniques to keep nutrients out of lakes. The 1,350-acre Roger Melear farm in Okeechobee was one of the first dairy farms to flush cattle manure from the milking barn into a reservoir where solids settle. The nutrient-rich water is used to irrigate pasture lands. This process keeps excess nutrients from flowing into nearby surface waters.

WHAT TO DO

- Use non-phosphorus detergents and soap products. Many detergents sold in the United States are either phosphate-free or low-phosphate. Check the labels.

- Don't bathe yourself, your dog or your children in the lake or wash a boat or car where the water will flow into the lake.

- Instead of using fertilizers, plant native plants and ground covers which need less attention.

- Keep septic tanks and filter beds in good operating condition. Check with your local public health agency for details.

- Use best management practices (BMPs) on agricultural lands near surface waters and recharge areas. The practices include spreading animal wastes away from waterways which reduces nutrients by 60 to 80 percent and placing animal feeders and fences away from waterways which reduces nutrients by 20 to 25 percent.

- Be a lake or river watcher. Anyone can be a waterway watcher. You can do it on your own or you can participate in the formal Lake Watch program sponsored by University of Florida Institute of Food and Agricultural Sciences. The state-funded citizen monitoring and education system involves over 1,000 members collecting water samples from over 300 lakes in 54 counties. Participants test water levels, clarity, algae content and total nitrogen and phosphorus. To start the program, watchers receive a sampling kit and four hours of training.

- Look for these 10 pollution warning signs.

    1. Dead or distressed fish, turtles, frogs, or other aquatic life.
    2. Oil slicks on the water surface.
    3. Frequent sores or tumors on fish.
    4. Bad odors in the water.
    5. Foul odors in the fish during filleting or cooking.
    6. Unusual algae blooms.
    7. Unusual water color or lasting muddy water.
    8. Accumulation of debris on the water surface or bottom.
    9. Accumulation of foam or bubbles that do not dissolve.
    10. Total absence of aquatic life over large areas of water.

Lakes
and
Rivers

RESOURCES

WATER MANAGEMENT DISTRICTS
(See appendix for a complete list.)

LAKEWATCH PROGRAM
University of Florida IFAS
7922 NW 71st St.
Gainesville, FL 32602
(904) 392-9617

***FLORIDA LAKES* AND *POLLUTION RECOVERY TRUST FUND* BROCHURES**
Florida Department of Environmental Regulation
2600 Blair Stone Rd.
Twin Towers Office Bldg.
Tallahassee, FL 32399-2400
(904) 488-4805

KEEP STORMWATER OUT OF WATERWAYS

- A 1989 survey showed 43 percent of Florida waterbodies analyzed were threatened and another 41 percent were suspected of being threatened by stormwater pollution.

- Some 2 million acres of Florida waters, including one million acres in the Everglades, have been placed under health alerts for fish consumption because of mercury contamination.

- Stormwater generates all the sediment and 80 to 95 percent of the heavy metals in Florida waterways.

- 82 lakes within the Orlando city limits have deteriorated due to stormwater run off.

- Police officers diving in the Miami River must wear state-of-the-art dive suits because the waters are so polluted the bacteria counts sometimes soar thousands of times above safe levels. The "dry" dive suits keep all skin from being exposed to the water.

- Indian River's clam industry once with 1,000 clammers and 150 dealers has been wiped out by pollution from sewage spills and stormwater run-off. The multi-million dollar industry has been diminished to 100 clammers and five dealers.

Lakes and Rivers

Stormwater is the largest polluter of Florida waters and the sole pollution source in some areas. Most of the direct sources of pollution such as industrial discharges are already regulated through permits. Insecticides, fertilizers, leaking sewage from septic tanks, and trace metals from the roadways are just some of the pollutants found in stormwater everywhere. In California, children are cautioned not to play in storm drainage ditches or swim in the ocean after a rain because of all the germs and pollutants washed into the stormwater. Most people think stormwater systems are primarily designed for flood control. Although that is an important reason, the water pollution problem is equally dangerous. Scientists are working diligently to design systems to filter the water of pollutants before it flows into our lakes and streams.

SUCCESS STORIES

- In the Biscayne Bay area of Miami, Florida, citizens are charged $2.50 per family per year in order to implement a complete stormwater system to help cleanup surface waters.

• Orlando has created the Greenwood Urban Wetland, an attractive park, from a stormwater retention area. Orlando's stormwater program is funded through the municipal utilities.

WHAT TO DO

• Protect rivers and lakes from erosion and stormwater drainage by planting sod or native plant barriers and using berms and swales.

• Put sod or gravel in place to control water runoff from drains.

• Limit the clearing of natural lakeside vegetation. Consider using a dock or moveable float to access swimming.

• Use porous materials for walks and driveways such as wood chips, gravel, leaves, pine needles, or the new porous paving materials to slow and filter stormwater run-off.

• Support the purchase of land corridors, floodplain property, or use of conservation easements along rivers and lakes.

RESOURCES

*CREATING A SUCCESSFUL NONPOINT SOURCE
PROGRAM: THE INNOVATIVE TOUCH*
Office of Water Regulation
Nonpoint Sources Branch
Washington, D.C. 20460

STORMWATER MANAGEMENT
Department of Environmental Regulation
2600 Blairstone Rd.
Tallahassee, FL 32301
(904) 488-6221

BE AN ECO-FRIENDLY FISHERMAN

Florida leads the nation in number of days of fishing by recreational anglers: 80.8 million days compared to California's 68.2 million days. The anglers spend more than $2.5 billion annually on recreational fishing.

WHAT TO DO

The catch-and-release approach to fishing is rapidly grabbing hold. The practice insures good catches for years to come and gives the angler an opportunity to tell great fish stories. If planning to catch and release use the following guidelines:

- Bring the right equipment including a hook remover or needle-nose pliers.

- Push down barbs on the hook before removing to make it easier to free the fish.

- Handle fish gently and keep in water as long as possible.

- For anglers that choose to keep some of their fish:

  Keep only what you and your family will eat.

  Do not kill and discard "trash" fish or non-edible fish. They still provide an important role in the ecosystem. They may be food for the big fish you hope to catch in the future.

  Obey the size limits and rules on certain species.

- Other points to remember:

  Use hooks and leaders that easily corrode so they will dissolve in the water.

  Don't cast when a bird is overhead.

  Pick up trash you see along the way. Plastic bags look like jelly fish and styrofoam looks like bread to hungry fish.

  If you see someone violating size limits or other fishing regulations, call 800-DIAL-FMP.

Lakes and Rivers

RESOURCES

FLORIDA LEAGUE OF ANGLERS
120 S. Olive Ave., Suite 705
West Palm Beach, FL 33401

ORGANIZED FISHERMAN OF FLORIDA
P.O. Box 740
Melbourne, FL 32902
(407) 725-5212

AN ANGLER'S GUIDE TO ENVIRONMENTAL ACTION
Anglers for Clean Water, Inc.
5845 Carmichael Rd.
Montgomery, AL 36117
(205) 272-9530

A free brochure on how fisherman can help protect our natural resources.

FLORIDA RECREATION SALT WATER FISHING FACTS
Florida Marine Patrol
Marjorie Stoneman Douglas Building
3900 Commonwealth Blvd.
Tallahassee, FL 32399
(904) 488-5757

List of regulations regarding size limits and protected species of fish and shellfish.

GAME AND FRESH WATER FISH COMMISSION
Farris Bryant Building
620 South Meridian
Tallahassee, FL 32399-1600
(904) 488-2975

Information of freshwater fishing rules and regulations.

FISH AMERICA FOUNDATION
1010 Massachusetts Ave., NW, Suite 320
Washington, D.C. 20001
(202) 898-0869

Provides funding to civic and angling groups focused on improving water and fishery resources.

FLORIDA CONSERVATION ASSOCIATION
904 East Park Ave.
Tallahassee, FL 32301
(904) 681-0438

MARINE FISHERIES COMMISSION
2540 Executive Center Circle, West
Suite 106
Tallahassee, FL 32301
(904) 487-0554

RIVERS WERE
MEANT TO BE FREE

"We can continue to alter the shorelines and essentially love the rivers to death, or we can back away, build our houses a little farther back and still enjoy the rivers but not be on top of them."

—*Michael Perry*
Southwest Florida Water
Management District

- Florida irrigates more than all the eastern states combined. Forty-seven percent of the water taken from rivers is used for irrigation. Eighty-one percent of the water is not returned.

- Before 1980, the Army Corp of Engineers had spent nearly $800 million on navigation and flood control projects in Florida.

- The straightening of the Kissimmee River resulted in the loss of 95 percent of the river's wildlife habitat and the ecological and hydrological destruction of Lake Okeechobee. The project destroyed 34,000 acres of wetlands and 14,000 acres of marshes. Restoration taking 15 to 20 years is going to cost approximately $423 million.

- Nine thousand acres of forest lands were destroyed along the Oklawaha River when the Rodman Dam was built, part of the defunct Cross Florida Barge Canal.

- The costs of keeping up man-made dams and reservoirs is high. For example, maintaining the Rodman Dam and reservoir costs taxpayers more than $500,000 a year.

- By 1980, 50,000 dams of 25 feet or higher had been built in the United States. By 1983, the Environmental Policy Center had stopped the building of 140 dams and canals, saving $20 billion and great ecological losses.

Lakes and Rivers

Florida's rivers are at risk. We have built our homes and parking lots too close. For years, we have dammed and diked our wild rivers, altering their natural flow and causing great damage to natural habitats.

The trend towards taming our wild and scenic waterways began back in the 1800s. In 1850, The Swamp Act transferred all wet and overflowing lands to the state. In those years, swamps were seen as useless, mosquito-infested lands. In 1865, the dredge, the mechanical tool that rid Florida of much of our wetlands, came into use.

Today, South Florida has a huge man-made water control network made up of 1,400 miles of canals, dams and floodgates.

During the 1960s and '70s, people began to realize the mistakes made nationally as river after river had been diked and dammed to create mud-ringed, monotonous, lifeless lakes and channels. Dams or reservoirs limit the movement of fresh water and destroy estuaries. Salt water begins moving further up freshwater streams, changing the types of fish and vegetation that normally thrive there.

The side effects of dams and canals are many. Fish migration often stops, reducing fish species such as redfish and snook. Ecosystems or unique habitats are altered or destroyed.

Sometimes we have to learn the hard way that it is hard to improve on a natural system.

SUCCESS STORIES

- Since 1981, Florida has acquired 447,236 acres of environmentally sensitive lands through the Save Our Rivers Program.

- The U.S. Army Corps of Engineers has approved the $423 million restoration plan for the Kissimmee River. The federal government has committed $127 million in restoration costs.

- States have begun to see the importance of restoring natural river systems. In Minnesota, 1,000 dams have been washed out or destroyed.

- The National Wild and Scenic Rivers program has brought recognition to these American treasures. In 1985, The Loxahatchee River was designated the first wild and scenic river in Florida.

- St. John's River Water Management District is implementing a major restoration project at the headwaters of the St. John's River.

WHAT TO DO

- Build a strong local group committed to a nearby river.

- Gather good information about the river such as possible pollution sources, habitat studies, usage figures and research that can help in developing a good river-management and education program

- Network with strong national organizations. Learn how to gain greater protection of Florida's rivers through effective legislation.

- Set up a consistent "river watch" program to keep a pulse on water conditions and look for new pollution sources.

- Educate the public on how to protect their natural river systems.

- Monitor local government comprehensive plans, new development orders for construction projects and other issues that may affect your local rivers.

RESOURCES

SAVE OUR RIVERS PROGRAM
  Department of Environmental Regulation
  2600 Blair Stone Rd.
  Twin Towers Office Building
  Tallahassee, FL 32399-2400
  (904) 488-0890

IZAAK WALTON LEAGUE
  4512A 86th St. West
  Bradenton, FL 33529
  (813) 792-8482

  Volunteer programs: Save our Streams and Wetland Watch. Support and provide brochures on protection of America's land, water, and air.

FLORIDA DEFENDERS OF THE ENVIRONMENT
  2606 NW 6th St.
  Gainesville, FL 32609
  (904) 372-6965

  Several active river protection and restoration projects.

ENDANGERED RIVERS
  by Tim Palmer

  Discusses the history of America's rivers and recent efforts to protect them.

AMERICAN RIVERS CONSERVATION COUNCIL
  2545 Blair Stone Pines Dr.
  Tallahassee, FL 30301
  (904) 942-0414 or 224-7896

  Worked with numerous agencies to review 50 of Florida's most important rivers.

Lakes
and
Rivers

THE COASTS

Florida has more coastline than any other state except Alaska—approximately 11,000 miles. Florida is also the largest "ocean-owning" state with 6.7 million acres of offshore land and 3 million acres of estuaries, open water, and wetlands under state ownership.

The coasts are very important to the economy of the sunshine state. Tourists list our coasts as the state's number one attraction. Our combined recreational and commercial fishing revenues of more than $7 billion annually makes us the most valuable fishery in the nation.

WITH THE MOST

    Beaches are so vital to Florida's economic and environmental future that the Florida Department of Environmental Regulation has many projects focusing on coastal issues. From developing more boat-mooring buoy sites in the Florida Keys (to save the reefs from anchors) to producing a sea turtle nesting map to finalizing a post-disaster plan for beach communities, the state is committed to saving the Florida coasts and their fragile ecosystems. In this chapter, you will learn ways you can help to protect Florida's coasts.

THE HIDDEN DUMPING GROUNDS

- Nationally, hundreds of millions of gallons of municipal sewage effluent (sewage that is slightly treated) are dumped into the ocean each day.

- One in four boaters admit to illegally discharging sewage into surface waters.

- Scientists have documented a 3,000-square-mile dead zone in the Gulf of Mexico where marine life cannot survive.

The *World Atlas* shows long stretches of Florida's coasts with "persistent coastal pollution" (long-term, but not necessarily severe). Ironically, the California coast, where swimming is often banned after rainstorms, is not as polluted as the Florida coast.

According to the *World Atlas* coast pollution is caused by the following:

44 percent direct discharge or discharge via rivers
33 percent fallout from air pollution
12 percent shipping
10 percent deliberate dumping
10 percent offshore oil and gas

WHAT TO DO

- Report boaters who discharge their sewage-holding tanks into rivers and oceans, breaking federal laws. Alert the local Coast Guard office or marine patrol.

- Support installation of dockside pumping stations along the coast where recreational boaters can empty their sewage-holding tanks.

- Avoid using bleach or formaldehyde-based compounds to clean boat heads. Instead, dump regularly to keep them clean.

RESOURCES

FLORIDA OFFICE FOR AMERICAN LITTORAL SOCIETY
Susan Holderman
New College/University of South Florida
5700 Tamiami Trail
Sarasota, FL 34234
(813) 351-3886

PROJECT REEFKEEPER
16345 W. Dixie Highway, Suite 1121
Miami, FL 33160
(305) 945-4645

THE CENTER FOR MARINE CONSERVATION
1725 DeSales St. N.W.
Washington, D.C. 20036
800-CMC-CLEAN

THE OCEANIC SOCIETY
218 D. St. S.E.
Washington, D.C. 20036
(202) 544-2600

Active projects related to ocean dumping, spills and contamination of the oceans.

U.S. COAST GUARD
2100 Second St., S.W.
Washington, D.C. 20593
(202) 267-2390

FLORIDA MARINE PATROL
Dept. of Natural Resources
3900 Commonwealth Blvd.
Tallahassee, FL 32303
800-DIAL-FMP

CITIZENS COASTAL ADVISORY COMMITTEE
Department of Community Affairs
2740 Centerview Dr.
Tallahassee, FL 32399
(904) 488-8466

Part of an extensive federal/state coastal management program.

WORLD WILDLIFE FEDERATION'S ATLAS OF THE ENVIRONMENT
by Geoffrey Lean, Don Hinrichsen, and Adam Markham.

The
Coasts
with
the
Most

THE COASTS GET SLIMED

- The coasts of Dade, Broward and Palm Beach counties experienced a huge algae bloom in the last three years. Some reefs had green, slimy algae 50 to 100 feet deep. This jello-like algae called Codium killed some coral reefs and threatened Florida's seafood industry. The algae blocks out sunlight, uses up oxygen and displaces fish and marine life.

- The red tides, occurring since the 1800s, may be a natural occurrence. The red tide blooms create nine neuro-toxins which kill fish creating a smelly stench which causes breathing discomfort in humans. Shellfish plagued by the red tide are poisonous. The red tides usually occurs between September and February.

- Hairlike algae cover Shark Reef in the Florida Keys smothering the multi-colored coral. The 4,000-square-yard algae bloom has destroyed an area that once was filled with shrimp, sponges and fish.

- Algae-fighting sea urchins were wiped out in the Florida Keys during 1983. Experts believe a viral disease may have killed them.

- Forty-two percent of the nation's waters are too polluted to allow the harvest of mollusks or shellfish.

From green slime to red tides, Florida's coasts have experienced some strange occurrences in recent years. Experts are at odds over the cause. Some say the phenomena are caused by natural or cyclical changes in climate. Others blame the weird and unsightly changes on polluted stormwater, septic-tank drainage, agricultural and industrial waste water or ocean "outfall" sewage pipes.

SUCCESS STORIES

- Hotels, such as the Cheeca Lodge in Islamorada, Florida, have implemented a wide variety of environmental programs to keep waste out of the waters of the Florida Keys. They use biodegradable, low-phosphate cleaning agents, drought resistant plants to conserve water and cut down on fertilizer, and grey water (treated waste water) to irrigate. (The Cheeca Lodge won the 1990 Gold Key Environmental Achievement Award from the American Hotel and Motel Association.)

WHAT TO DO

WHAT TO DO

- Call for enforcement of the Federal Clean Water Act which mandates zero discharge of toxic pollutants in waterways.

- Eliminate use of toxic chemicals, especially near coastal waters. Collect marine paints, varnish, boat oil and other toxic items to deposit at a toxic waste collection center.

- Reduce high-nutrient and contaminated run-off. Use drought-tolerant native plants around coastal waters to reduce the need for fertilizing. Check stormwater drains to make sure that illegally treated wastes are not being dumped directly into the ocean.

RESOURCES

FLORIDA DEPARTMENT OF ENVIRONMENTAL REGULATION
Coastal Zone Management
2600 Blair Stone Rd.
Twin Towers Office Building
Tallahassee, FL 32399-2400
(904) 488-0890

TAKE BACK THE COAST
218 D. St. S.E.
Washington, D.C. 20002
(202) 544-2600

COAST ALLIANCE
1536 16th St. N.W.
Washington, D.C. 20036
(202) 265-5518

The
Coasts
with
the
Most

THE THREAT OF OFFSHORE OIL

*"Drilling in fragile, environmentally sensitive areas
will not build a sound energy policy and will not
solve the immediate oil crisis."*
—*Senator Bob Graham*
August 19, 1990
Orlando Sentinel

- Each year, 5,000 oil tankers pass through Florida waters.

- Reef groundings caused $20 million in damage during spring and summer 1990 alone. The new U.S. Marine Sanctuary designation restricts large ocean- going vessels from traveling inside the sanctuary boundaries.

- During the 1980s, the federal government issued 73 offshore oil leases in the southern Gulf of Mexico. None has been drilled as yet. They may one day be used to supply U.S. oil needs.

- In June of 1991, 1,000 gallons of oil seeped from a faulty valve on an underwater pipeline in Tampa Bay. The spill threatened birds, fish, and shellfish.

Offshore oil drilling and transportation still remain a threat to Florida's coasts. Although President Bush delayed oil drilling in the Florida Keys until year 2000, the measure was not permanent or comprehensive enough. And, Keys advocates are pushing for stronger laws regarding oil tanker movement near the magnificent coral reefs.

Florida's coastal treasures—such as its mangrove swamps, seagrass beds, coral reefs and barrier islands—are not the only resources that could be damaged. The Alaskan oil spill caused losses of $110 million to regional fisheries and $260 million in economic associated activities.

Some proponents say offshore oil drilling carries less risk than the greater danger of importing oil from other countries. But, the long-term answer to our energy problem does not lie under the U.S. continental shelf. Excluding the central and western Gulf of Mexico, U.S. offshore oil represents only 323 days worth of energy at 17 million barrels per day. Offshore oil does more than damage the marine environment, it threatens Florida's shore communities as well. It has the potential to degrade water quality, add additional air pollution, create noise pollution from boats and airplanes and damage wetlands through the laying of pipelines and pipeline leaks.

The controversy will continue as environmentalists weigh the trade-offs and ongoing studies are completed. Not enough research has been done on the effects of an oil spill near the Keys and southern coast of Florida. Current research does show damage in warm water is worse than previously thought. Off the coast of Panama, an oil spill killed submerged coral reefs, when oil "rained down" on the reefs.

Many see offshore drilling as America's answer to becoming energy self-sufficient.

WHAT TO DO

• Revive the energy-saving ethic of the 1970s. Buy a smaller automobile and use alternative energy sources when possible. (See "Global Warming," page 129 and "Sustainable Energy," page 134.)

• Lobby Congress for stricter rules on transportation of oil in and near the Florida Keys sanctuary.

• Discourage offshore oil drilling along the Florida coast until comprehensive studies have been completed.

RESOURCES

FLORIDA PUBLIC INTEREST RESEARCH GROUP
308 East Park Ave., Suite 213
Tallahassee, FL 32301
(904) 224-5304

HOOVER ENVIRONMENTAL GROUP
7420 S.W. 59th Ave.
Miami, FL 33143
(305) 665-6369

THE OCEANS, OUR LAST RESOURCE
by Wesley Marx

The
Coasts
with
the
Most

DISAPPEARING SANDS

- Florida has more than 300 miles of critically eroding beaches.

- According to the Florida Department of Natural Resources estimates, 600,000 cubic yards of sand flow past the Fernandina Inlet, north of Jacksonville, each year.

- In Florida, approximately $23 million in state, federal and local monies were spent in 1991 to renourish state beaches. In 1990, more than $46 million was spent. Renourishing must be repeated every seven to 10 years.

- Florida's beach-armoring policy (restricting the use of walls and other artificial barriers) is far behind other southern states. No new beach-armoring is allowed in North or South Carolina.

Beaches constantly shift because of tides, currents and hurricanes. No one can stop these forces.

Left to nature, beaches will erode during a storm and rebuild afterwards. Waves carry the new sand down-current. To try to control the natural process with the construction of artificial seawalls, jetties and other obstructions causes greater beach erosion in the long term. Jetties trap sand and keep it from moving down-current to the next beach.

Seawalls, or hard structures used to ward off wave energy, force the waves down and cause erosion under the structure. Beaches with jetties and seawalls cannot rebuild themselves.

In addition to causing erosion, seawalls or armoring prevent sea turtles and other wildlife from using the beach. In fact, the beaches most likely to be eligible for armoring permits are the same areas where most sea turtles nest.

The natural way to prevent erosion is with dunes. Dunes keep sand from blowing away or washing away with a storm. This is why dunes are so precious to Florida's beach communities. Removing sea oats which provide dune protection is against the law and destroying dunes with off-road vehicles is prohibited. In some Florida coastal communities, development setbacks are used to prevent dunes from being destroyed by construction.

WHAT TO DO

- Don't damage the sand dunes by picking sea oats or other natural vegetation.

- Use boardwalks to reach the beach to keep from damaging the dunes.

- Lobby against the use of beach armoring, seawalls or other artificial structures.

RESOURCE

CENTER FOR MARINE CONSERVATION
Florida State Office
One Beach Drive, S.E., #304
St. Petersburg, FL 33701
(813) 895-2188

The Coasts with the Most

BE A GOOD SPORT AROUND FLORIDA'S COASTS

- Soft corals, or Octorals, contain colonies of many small animals that are important to the food chain. Not all Octorals are protected. More than 20,000 are collected a year creating a $15-million state industry. They are often used in the production of prescription drugs.

- An estimated 300 tons of "live rock" is hauled annually from around the Florida Keys and the Tampa Bay areas to be used in salt water aquariums. The "live rocks" are home to hundreds of tiny animals and larvae. The practice is chipping away part of Florida's undersea communities.

- Of the two million annual visitors to the Florida Keys, one million snorkel or scuba dive.

- The United States has more than four million scuba divers and another 10 million are anticipated to become scuba divers within the next 10 years.

Millions of visitors flood Florida's beaches annually. Americans' love of the great outdoors can put stress on the coastal environment. Even some of the most subtle sports, such as shell collecting, sailboarding and snorkeling, can be destructive. But water sports enthusiasts do not have to harm the coast. By practicing a little "coastal courtesy" and remembering that they are a guest of the sea, visitors can help protect the environment.

WHAT TO DO

- Sponsor a coastal clean-up.
  - Educate others about the coastal environment.
- Donate to programs designed to help the coast.
- Leave the coast just as you found it.
- Report violators who dump pollutants into navigable waters (Violation of the Federal Water Pollution Control Act of 1972).

WHAT TO DO
FOR SEASHELL COLLECTORS

- Teach children about marine life and the food chain by pointing out the little creatures found living within shells. Encourage them not to take a shell or rock that is home to a live sea creature.

- Do not collect shells in excessive numbers. In the 1980s, the beautiful, edible pink conch was depleted by commercial fish-market suppliers and over-zealous collectors. The conch is now protected.

- Be a beach watcher. Check for unusual signs on the beach that may indicate pollution. Pick up trash that may be washed back into the ocean.

- Report people who may be illegally removing protected conch shells or polluting the coasts, to 800-DIAL-FMP.

WHAT TO DO
FOR DIVERS

- Be careful not to break off coral by carelessly hitting it with your diving fins or equipment. Coral takes years to grow. Excessive diving in some areas of the coast has left patches of algae-covered coral, broken sponges, and cracked coral branches.

- Check your buoyancy and do not carry too much lead.

- Do not handle sea creatures. It puts great stress on them.

- When photographing, do not alter the undersea landscape to get the perfect picture.

WHAT TO DO
FOR JET SKIERS

- If you like to Jet Ski, do not go into refuge areas. Scaring nesting birds and disturbing grass beds where young fish are born will endanger the wildlife the refuges are there to protect.

The
Coasts
with
the
Most

RESOURCES

FLORIDA MARINE PATROL
Department of Natural Resources
3900 Commonwealth Blvd.
Tallahassee, FL 32303
800-DIAL-FMP

THE GREAT BOAT BRIGADE

- The U.S. Coast Guard estimates that each boater on a pleasure boat generates a pound and a half of garbage a day.

- Nine out of every 100 people in Florida buy boats.

- Thirteen out of 37 coastal counties in Florida are implementing speed zones.

- Florida has the highest boating accident and fatality rate in the nation (1,176 accidents in 1990).

- One quart of boat oil can spread a suffocating two-acre slick on the water.

- It has been estimated that the oil inadvertently dumped in the world's coastal waters every year is slightly more than the amount spilled by the Exxon *Valdez* in Alaska.

Florida had 718,054 registered boats in 1991. According to Audubon Society Executive Director Charles Lee, boats collectively do more damage to the environment than large dredge and fill equipment. Boaters can enjoy Florida's lakes and waterways and help protect them.

SUCCESS STORY

- In Kings Bay, Florida, near Crystal River, local residents devised a program to grow native seagrass, a beneficial water plant. Residents pulled hydrilla out of the river and replaced it with sprigs of eel grass (Vallisneria). The native seagrass was grown from pieces torn from the bottom by motorboats. Native seagrass provides habitat for fish, absorbs nutrients, and is less likely than hydrilla to get caught in boat props.

WHAT TO DO

- Check your boat engines to make sure they are not leaking oil or gas. Inspect fuel lines for cracks, too.

- Stay away from shallow areas and seagrass beds to keep from stirring up sediment and harming plant and animal life. Lift motors when coming into shore and crossing grassbeds.

- Inspect your boat trailers and motors after loading to ensure that you are not carrying exotic aquatic plants to another lake or waterway.

- Don't litter or pollute the water. Put a trash bin with a locking lid on board.

- Drive slowly in manatee zones and always keep on the alert for surfacing marine animals such as sea turtles.

- In reef areas, use mooring buoys or drop your anchor in a sandy area to keep from damaging coral reefs.

- Install a propeller guard on your motor to protect manatees, turtles, and other marine life.

- Fill tanks precisely to keep overflow from spilling into the water.

- Buy boat oil in bulk and carry the extra oil in reusable containers. You'll save money and hundreds of plastic bottles.

- Curb and control use of hazardous boating materials such as marine paints, solvents, cleaners and petroleum products. Use materials on land and over a drop cloth. Take the drop cloth to a toxic disposal bin.

- Support a boat-driver license program which would include both safety and environmental awareness education.

- Consider a canoe. Fishermen can see things from a canoe that they would not normally see from a motor-boat. You can have a wilderness experience, sneak up on wildlife and get into hard-to-reach areas. A new fisherman's canoe has been designed by a Gainesville manufacturer. It has rod holders, nylon anchor cleats, a gravity operated boat well or live well, and a wider hull. It only weighs 90 pounds.

RESOURCE

FLORIDA MARINE PATROL
   Department of Natural Resources
   3900 Commonwealth Blvd.
   Tallahassee, FL 32303
   (904) 488-1234

The
Coasts
with
the
Most

COASTAL PARADISE: THE FLORIDA KEYS

Anyone who has ever visited the Florida Keys would probably agree with the Nature Conservancy that it is truly one of the "Last Great Places." Home of one of the world's largest coral reef systems, the island chain is one of 12 international sites chosen by the Nature Conservancy for special protection. Thousands of years ago the islands were cut off from the mainland by rising seas. One half of the Keys are covered with Mangrove swamps. The Keys also support the largest tropical hardwood forest in the United States. Although tourists are often surprised that there are no great white beaches in the Keys, they soon realize that the true magic lies underwater. A scuba or snorkel trip to the Key's living reefs makes you feel like you just stepped inside one of the most amazing tropical aquariums with more than 300 species of fish.

During l990, the U.S. Congress designated 2,600 square nautical miles of area off the Keys as a marine sanctuary. The Keys also boast four National Wildlife Refuges, a portion of the Everglades National Park, and numerous other public and private parks. With all the park lands, you might think that the Florida Keys would be protected from environmental destruction. Unfortunately, that is not the case. The Keys are endangered by overuse and water pollution.

The same natural beauty that qualified for park designations draws over two million tourists annually. A million of those guests dive in the Keys. Boat anchors scar, and divers damage the fragile coral reefs. Coral reefs are declining by four percent each year while this unique living community takes 30 years to grow merely one foot.

Although divers take a dramatic toll on this undersea city, it is what happens above ground that does the greatest damage. The Key's killers include water pollution from septic tanks, pesticides, dredging and filling, and other sources. Although cesspools were outlawed 30 years ago, there are still 5,000 to 8,000 active cesspools remaining in the Keys. That means raw sewage is run into the groundwater. The over 30,000 septic tanks create an enormous potential for sewage to leak through the porous soil and limestone. But one of the greatest sins of all is the raw sewage dumping from house boats and other marine vessels that legally are required to use holding tanks and pump out while dockside. Amazingly, until 1990, the community of Key West itself pumped untreated sewage less than a mile offshore. Now the city of Key West has the only sewage treatment facility in the entire chain of islands. The heavy nutrients from this sewage and agricultural runoff are literally killing the reefs. Algae blooms are widespread. The growth blocks out the sunlight and kills living organisms below.

Fish, shrimp, and lobster catches are down. The water clarity is poor.

The good news is that everyone is beginning to work together. People who make their living from tourism and fishing are realizing that the future of the Keys will depend on good environmental measures established today. Reef advocates hope that the federal governments plan to develop a better management program within two years will head off destruction. President Bush has stopped offshore leasing activity for at least 10 years. The federal government may ban shipping activity close to the reefs. Local "reef relievers" are cleaning up the ocean floor and trying to educate tourists coming into the area. Consideration is being given to restricting fishing and lobstering in some park areas. The South Florida Water Management District is working on stormwater solutions.

The Nature Conservancy does not think it is too late to rescue this ecological wonderland. With reef rescue in high gear, it appears they may be right.

RESOURCES

FLORIDA KEYS LAND AND SEA TRUST
    P.O. Box 536
    Marathon, FL 33050
    (305) 743-3900

THE NATURE CONSERVANCY
    P.O. Box 4958
    Key West, FL 33040
    (305) 296-3880

REEF RELIEF
    P.O. Box 470
    Key West, FL 33041
    (305) 294-3100

CORAL REEF COALITION
    Center for Marine Conservation
    1725 DeSalles St. N.W.
    Washington, D.C. 20036

MARINE AND ESTUARINE MANAGEMENT DIVISION
    Office of Ocean and Coastal Resources
    NOAA
    1825 Connecticut Avenue, N.W.
    Washington, D.C. 20235

The
Coasts
with
the
Most

LIVING AS IF THERE WERE A TOMORROW

"In our every deliberation, we must consider the impact of our decisions on the next seven generations."
—From the Great Law of the Hau de no saunee

Our ancestors once saw nature as infinite and inexhaustible. Their ignorance led to destruction that we can understand and forgive.

Today we can no longer say that "we didn't know." Some of the Earth's resources are finite, and others renewable only if we care for them. Our scientific and technical know-how make us capable of, and ethically responsible for, the preservation of our planet.

Indifference and despair are self-indulgences we cannot afford. In this chapter you will learn ways to preserve resources, save energy, and combat global warming and ozone depletion.

REDUCE, REUSE, AND RECYCLE, AND TRASH THE THROWAWAY SOCIETY

"Our economy is such that we 'cannot afford' to take care of things: Labor is expensive, time is expensive, but materials—the stuff of creation—are so cheap that we cannot afford to take care of them."

— *Wendell Berry*

- Floridians produce more than 16.5 million tons of solid waste annually. That's enough to pave every interstate roadway in the state with a layer more than four feet deep.

- In Florida, roughly eight pounds of garbage are thrown away daily per state resident. That's about twice the national average. Forty million annual visitors contribute to our high waste-generation rate.

- Ten percent of Florida's solid waste is recycled. West European countries recycle 30 percent. The rate for Japan is an impressive 50 percent.

- Fifteen million tires are discarded each year in Florida. That's more than one tire per person. If gathered into a pile 10 to 15 feet high, they would cover 40 acres of land.

- Floridians recycle 40 percent of aluminum cans, 10-15 percent of glass containers, and less than three percent of plastics used.

- Key West's abandoned dump, dubbed "Mount Trashmore," grew to 90 feet above sea level. The pyramid of garbage had risen as high as it could go without making the base broader, and sky-rocketing land prices made that impossible. Late in 1990 the city began trucking the trash 200 miles north to Broward County for disposal.

Florida's landfills are nearing capacity; many are overflowing, and some are contaminating our water supplies. While this is a problem most states are wrestling with, it's especially troublesome in our state, because we live and pile our garbage directly above our drinking water supplies.

When we practice the three Rs—Reducing, Reusing and Recycling

—we do more than slow the runaway growth of our dumps. We save the energy involved in excavating raw materials and manufacturing finished products; we save resources; we save land and wildlife habitat by reducing the deforestation of land to get to resources; and we reduce air pollution. That's why buying fewer things and reusing what we already have are even more important than recycling.

Florida law mandates that communities achieve a 30 percent reduction in waste disposal by the end of 1994. We can no longer afford not to take care of things.

SUCCESS STORIES

- Florida's Solid Waste Management Act is the most comprehensive in the U.S. Two million single family households—40 percent of all single family homes —participate in curbside recycling.

- Florida's Collier County Landfill Reclamation Project is recognized as one of the most creative and successful government projects in the nation. The Project's approach to garbage disposal has won it an award from the Ford Foundation and The John F. Kennedy School of Government at Harvard University. Collier County sees a landfill as a process by which solid waste decomposes into recyclable materials rather than as a place to store garbage. Older, unlined dumps are "mined" for ferrous metals, glass, aluminum and plastics. The recyclables are then sold while other reclaimed materials are used to cover landfills or as soil additives.

- Florida encourages the use of recycled newsprint by taxing virgin newsprint 10 cents per ton.

- The State Department of Transportation is field-testing recycled plastic fence posts instead of treated wooden posts. The recycled plastic is more durable, as well as being more environmentally sound.

- Florida has a tire-disposal program financed by fees levied on retail sales of new tires. One dollar is collected for each new tire sold. The fund is used to help counties pay for local tire disposal or recycling efforts, for state cleanup of old-tire dumps and for research into new uses for old tires.

- In Dade County, where 25 to 60 percent of the waste stream is yard waste, the Metro-Mulch campaign collects and shreds yard waste and distributes mulch free.

- Florida's hospitality industry generates 10 percent of the state's 19.3 million tons of solid waste annually. Six leading Orlando hotels participated in a pilot program designed to test the most

Living
As If
There
Were A
Tomorrow

efficient method of recycling waste. During the 10 month program 300 tons of solid waste were recycled, saving 15,000 cubic feet of landfill space. The participating hotels found they can save from $300 to $3,000 per month by recycling.

WHAT TO DO
REDUCE

• Take your own canvas or string bags to the store and say "no" to paper and plastic.

• Purchase food in bulk, or largest unit available to avoid excess packaging.

• When products are only offered wrapped individually, buy those in recycled and recyclable packages.

• Use rags instead of paper towels. Rags cut from old clothes, sheets or towels are free, and can be washed with your laundry.

• Carry your own mug for coffee takeouts.

• Cover bowls of leftovers with shower caps instead of plastic wrap. Wash them with your dishes between uses.

• Make or buy cloth napkins instead of buying paper.

• Use a litterless lunch box or reusable cloth bag instead of disposable lunch bags.

• Buy rechargeable batteries and a recharger. You'll avoid sending batteries to the landfill where the resulting release of heavy metals, such as mercury and lead, could contaminate your groundwater.

• If you shop by mail, write a note on the order form asking that biodegradable packing material be used.

• Use a gold mesh coffee filter that can be rinsed between uses and lasts indefinitely, instead of buying paper filters. Or try the less expensive reusable cotton filters that will last a year or more. If you must use paper, buy the unbleached kind. (Bleached paper filters have been banned in Sweden because hot water can leach dioxins into the coffee.)

• When you shop, pick things made to last instead of disposables. Longer-lasting items may cost more initially, but turn out to cost less in the long run.

• Don't buy juice boxes or other disposables that can't be recycled.

WHAT TO DO
REUSE

- Buy used items when practical.
- Take time to maintain and repair the things you have.
- Use cold- and gentle-washer settings to make your clothes last longer.
- When you can no longer use something, give it to a friend or charity, or save it for a yard sale.
- Reuse large glass jars to store dried beans, rice, macaroni and other bulk items in the kitchen.
- Use small jars for storing leftovers in the refrigerator or nails and screws in the shop.
- Reuse wrapping paper, plastic bags, boxes and lumber.
- Stick labels over addresses on unused return envelopes, and use them when you pay bills by mail, or for informal correspondence.
- Reuse yard waste such as tree leaves, pine needles, and grass clippings, as mulch.

WHAT TO DO
RECYCLE

- Participate in recycling programs in your community.
- Wash your recyclables in your already dirty dishwater rather than using extra soap and water. They don't have to be spotless, just clean enough not to attract bugs.
- Encourage your office or workplace manager to place a recycling bin next to the soft-drink vending machine.
- When it's time to replace your car battery, insist on leaving your old one where you buy the new one. Florida law requires anyone who sells batteries to accept old ones to be recycled.
- When you buy new tires, trade in your old ones. If they're too worn for retreading, look for other ways to use them. They make great swings, planters, or boat bumpers. Slit and placed around young trees, they offer tender trunks protection from weed-eaters. In New Mexico and Colorado solar

Living
As If
There
Were A
Tomorrow

powered houses are made of them! (see Resources for more information.)

- Recycle books and magazines by donating them to public libraries, schools, or nursing homes.

- Turn kitchen scraps and grass clippings into compost. Apartment-dwellers can create compost in a small garbage can in the kitchen, without attracting flies or producing unpleasant odors! (See Resource section.)

- If you change your own oil, collect it in a container and take it to a recycling center. It's not worn out, it's just dirty. Most of Florida's 67 counties have collection centers for used oil. Consumer-accessible, used-oil collection igloos are being located at some convenience store sites by a Texaco affiliate, Star Enterprises. Your county information center can direct you to the nearest used-oil collection center.

- Take leftover products such as drain cleaners, toilet-bowl cleaners, and oven cleaners to a hazardous-waste collection site or licensed disposal facility.

- Purchase products made of recycled materials. Reliable markets for secondary materials are crucial to the success of recycling programs.

RESOURCES

FOR MORE INFORMATION ON RECYCLING IN YOUR AREA, CALL:

Florida Department of Environmental Regulation,
(904) 922-6104

Keep Florida Beautiful, Inc.,
(904) 561-0700

Florida Retail Federation,
(904) 222-4082

REDUCE—REUSE—RECYCLE (HE 3157)
Institute of Food and Agricultural Sciences
University of Florida, Gainesville
Available from your county extension agent.

CITY SAFARIS
by Carolyn Shaffer and Erica Fielder
Provides details on creating "garbage-can compost," and other topics of interest to urbanites of all ages.

EARTHSHIP, VOL. 1
by Michael Reynolds
Survival Architecture
P. O. Box 1041
Taos, NM 87571
Explains how energy-efficient, solar houses are built with tires.

TIRE RECYCLING INSTITUTE
c/o Ridgway Land Company
Drawer 500 / Ridgway, CO. 81432
(303) 626-5805

The Institute is in the process of formation as this book goes to press. Its purpose is "to educate the public and the elements of commerce and industry involved in the manufacture, sale, use and repair of tires on the subjects of responsible reduction, reuse, recycling and recovery of tires." Inquiries are welcome.

BETTER THAN WOOD
2000 S.W. 31 Ave.
Pembroke Pines, FL 33009
(305) 962-2100

Produces substitutes for wood, metal and concrete out of recycled plastic.

Living
As If
There
Were A
Tomorrow

DON'T LET FLORIDA GET BURNED BY GLOBAL WARMING

- According to scientists' estimates, the projected one-foot rise in sea level in the next few decades will erode between 300 and 400 feet of Florida's beaches. That represents the loss of more than $100 billion in investments, in addition to the loss of valuable ecosystems.

- A mid-range estimate of sea-level rise due to thermal expansion of the ocean is roughly two feet within the next century. An increase of only half that magnitude could amount to up to 200 feet of beach erosion in Florida.

- Many scientists believe that episodes of coral bleaching already occurring in Florida reefs (as well as in some locations in the Caribbean and the Bahamas) are the result of elevated sea-surface temperatures.

- According to estimates made by the Environmental Protection Agency, we can expect 200,000 additional deaths from skin cancer during the next 50 years due to the depletion of stratospheric ozone over the United States.

- In addition to causing skin cancer, the ultraviolet radiation that reaches us through the thinning ozone layer suppresses our immune systems, fostering the growth of other cancers. It has also been identified as a major cause of cataracts.

- Ground-level ozone is a potential threat to Florida's multi-billion-dollar agriculture and forest industries. The decline of the shade-tobacco industry in Florida is blamed by some experts on ozone pollution.

- On a hot summer's day in the greater Los Angeles area, emissions from charcoal lighter fluid can add up to four tons of reactive hydrocarbons. That's a bit more than twice as much hydrocarbon pollution than is produced by an oil refinery on a given day.

Florida's geographic location and topography make it particularly vulnerable to the effects of global warming. If the predictions of many scientists come to pass, the rise in sea levels will flood 30 percent of the

Living
As If
There
Were A
Tomorrow

state, and salt water will intrude into the fresh groundwater we depend upon for drinking water. The remaining land will be buffeted by stronger hurricanes, fueled by the rising temperatures. And as the protective ozone layer surrounding the earth is depleted by the gasses associated with global warming, skin cancer could become endemic and food crops could fail.

Carbon dioxide is the gas that makes the largest (49 percent) contribution to global warming. Nitrous oxide is another, though its contribution (six percent) is much smaller. What they have in common is that they both come primarily from the burning of fossil fuels and deforestation. These are things we can do something about.

Chlorofluorocarbons (CFCs), used for refrigeration, air conditioning, aerosols, foam blowing, and solvents, also play a part (15 percent). Again, we can exercise some control over their use.

Methane, too, has a role (18 percent) in the greenhouse effect, and is possibly the gas we as individuals can do the least about. Its sources are cattle, decaying organic matter, rice paddies, gas leaks, mining, and termites.

The last greenhouse gas (12 percent) we'll discuss here is ozone, which is formed by the action of sunlight on chemicals in the air. Ozone has a split personality. It's a pollutant near the ground where it helps form smog and acid rain. But at 12 to 30 miles up in the sky it forms a protective layer, limiting the amount of harmful ultraviolet radiation that reaches Earth. It can be confusing, but the important thing to remember is that the same actions that limit the production of "bad" ozone, also help protect the "good" ozone.

Though the study of climate change is young, if we wait until detailed projections can be made, it will be too late to deal with the threat of rising seas and shifting climate zones. We already invest money and time in defense programs to protect us against potentially devastating but uncertain risks. In the same way, actions we take to slow or avert global warming can be considered "insurance." Many of the steps we can take offer the additional benefit of saving money.

> *"After all, what's the use of having developed a science well enough to make predictions, if in the end all we're willing to do is stand around and wait for them to come true."*
>
> —Sherwood Rowland, *who sounded an alarm about ozone holes in the '70s*

SUCCESS STORIES

- Florida is the fourth state to join the Environmental Protection Agency's Green Lights program, a nationwide effort to get companies and state and local governments to invest in new lighting technology. By upgrading the lighting in state office buildings, the state expects to be saving 210,000 kilowatt hours (and $78 million in state utility bills) each year by the end of the five-year conversion program. That translates to a 320-million-pound reduction of carbon dioxide emissions from power plants.

- Florida law requires garages or automobile salvage operations to recycle Freon.

- A bill passed by the Florida legislature in 1987 introduced stringent efficiency standards for refrigerators and freezers, and added "second-tier" standards for fluorescent light ballasts and low-flow shower heads among others, starting in 1993. It is expected that consumers and businesses will save $4 billion in energy costs between 1988 and 2016.

- Kindergartners and first graders in Florida public schools benefit from an energy and conservation curriculum designed by the Governor's Energy Efficiency Office and implemented by the Florida Federation of Women's Clubs.

WHAT TO DO

- Drive the most fuel-efficient car available. It's one of the most important steps you can take as an individual. The vehicles we drive generate more pollution than any other single human activity.

- Use unleaded gasoline.

- Car-pool when possible.

- Check your tire pressure monthly. The government estimates that under-inflated tires cost the nation more than 100,000 barrels of oil a day.

- Ride your bike for five miles worth of traveling each week. If every driver in the country did, we could keep an estimated 6.7 million tons of carbon dioxide out of the air annually.

- Turn off your engine rather than letting it idle for more than one minute.

- Buy Florida-grown fruits and vegetables to eliminate the energy use associated with shipping produce from distant growers. The food on the average American dinner plate has traveled 1,300 miles.

Living
As If
There
Were A
Tomorrow

- A side-by-side refrigerator-freezer consumes 35 percent more energy than a model with the freezer on top. When it's time to replace it, choose one that uses less energy.

- Turn your water heater thermostat down 10 degrees and save six percent in energy costs. Water heating consumes 15 to 25 percent of all home energy.

- Insulate your hot-water heater to cut down on heat loss. Blankets or jackets made for the purpose cost about $20 and are available at stores that sell plumbing supplies or building materials.

- Lower the thermostat on your hot water heater when you plan to be away for a few days, or turn it off.

- Keep your air conditioning and heater thermostats set as close to outside temperatures as your comfort will allow. You can save up to 30 percent of the energy required for cooling and heating for each six degrees that indoor temperatures are brought closer to outside temperatures.

- Replace incandescent light bulbs with compact flourescents. Don't the let $15 plus price of the compact flourescent fool you. The bulb will last 10 times as long as an incandescent one, and uses only a quarter of the energy to produce the same light. Over the lifetime of the bulb you can save $50 in electricity.

- Use an electric starter, pine cones dipped in paraffin, or newspaper under the coals to start your barbecue instead of lighter fluid.

- Support alternative energy programs.

- Urge your representatives to support legis-lation promoting energy efficiency, renewable energy resources and a ban on the use and production of CFCs (chlo-rofluorocarbons).

- Avoid buying CFC-containing products. That includes foam cushions and mattresses.

- Plant trees. They help cool cities, shade homes, and absorb carbon dioxide.

- Recycle. It saves energy.

RESOURCES

THE GREENHOUSE TRAP, What We're Doing to the Atmosphere and How We Can Slow Global Warming
by Francesca Lyman

THE NEXT ONE HUNDRED YEARS, Shaping the Fate of
Our Living Earth
by Jonathan Weiner

U.S. DEPARTMENT OF ENERGY CONSERVATION AND RENEWABLE ENERGY
INQUIRY AND REFERRAL SERVICE
800-523-2929
Ask for their fact sheet on energy-saving measures you can adopt at home.

NATIONAL ENERGY FOUNDATION
5160 Wiley Post Way, Suite 200
Salt Lake City, UT 84116
(801) 539-1406
Offers a catalogue of Resources for Education.

GOVERNOR'S ENERGY EFFICIENCY OFFICE
214 S. Bronough St.
Tallahassee, FL 32301
(904) 488-2475

Resources Available from Your County Extension Agent, Produced by the Institute of Food and Agricultural Services at the University of Florida:

A GUIDE TO RESIDENTIAL ENERGY EFFICIENCY IN FLORIDA (EES 7), SAVING
ENERGY DOLLARS IN YOUR CONDOMINIUM THROUGH COST AVOIDANCE
(EES-27)

GLOBAL CLIMATIC CHANGE PRIMER (EES 72)

Mail Order Sources for Energy Saving Products:

REAL GOODS TRADING COMPANY
966 Mazzoni St.
Ukiah, CA 95482
800-762-9214
FAX: 707-468-0301

SAVEENERGY COMPANY
2410 Harrison St.
San Francisco, CA 94110
800-326-2120

SEVENTH GENERATION
Colchester, VT 05446-1672
800-456-1177
FAX: 1-800-456-1139

Living
As If
There
Were A
Tomorrow

SUSTAINABLE ENERGY IN THE SUNSHINE STATE

- Before World War II, Florida was a leader in the use of solar hot-water systems. In the 1930s almost half of the new homes built in Florida had solar-heating water systems. These systems lost their popularity when electrical energy became cheap. Solar heating of water for pools and residential use remains the most common application of solar energy in the state.

- About 500 million kilowatt hours of electricity are being saved each year in Florida through the use of approximately 250,000 domestic solar water heaters.

- Roughly 100,000 solar pool heaters are installed in Florida with savings amounting to about 52,000 cubic feet of natural gas annually.

- The University of South Florida has been chosen by the U.S. Department of Energy to test electric vehicles. The project involves the use of 200 solar panels mounted atop covered parking spaces. Such a facility could charge commuters' solar-powered cars during the work or school day.

- Solar-powered water pumps for cattle work best when they're needed most—hot, dry, sunny weather. They're being put to the test by the S.W. Agricultural Research and Educational Center in Immokalee. They are expected to be used more widely in the future when cattle may not be allowed to wander on wetlands.

- A fleet of 35 vehicles operated by Neighborly Senior Services in Clearwater is powered by compressed natural gas (CNG).

- A fleet of Gainesville city buses and trucks are being retrofitted for CNG by a grant from the Florida Energy Office and a Texas-based subsidiary of Transco Energy, Inc.

- Florida's first retail CNG refueling operation opened in the city of Sunrise in 1991, with more openings expected in 1992. This and other alternative-fuel initiatives got a boost when Governor Lawton Chiles signed an executive order requiring state agencies to begin planning to phase in "the least-polluting, alternative fueled" vehicles in areas of high pollution.

Florida's economy, largely dependent on tourism and agriculture, is highly energy intensive. Because we must import nearly all of the fossil fuels we use, it's especially important that we make use of renewable energy sources. Nowhere is this more critical than in the transportation sector, which accounts for about 36 percent of the state's total energy consumption and 75 percent of our petroleum consumption.

The sun, hydrogen, and compressed natural gas are among the viable alternative energy sources in Florida's future.

PHOTOVOLTAICS
(SOLAR ELECTRICITY)

Photovoltaics has proven to be feasible and reliable through the space program's use of solar cells to power satellites. By generating electricity directly from sunlight, they eliminate the need for fuel, and no noise or harmful emissions are produced. Currently about 400,000 kilowatt hours of electricity are produced annually in Florida using photovoltaics, for use in utility installations, state-funded demonstrations, and residential homes.

In our state, known for its sunshine, solar energy has tremendous potential. Sunshine is not only free, it's more widely distributed than any other energy source; its energy can be captured with only minor modifications in building construction, design or orientation; and it's especially well suited to supplying heat at or below the 220°F boiling point of water, making it a natural for cooking and heating. It could go a long way toward reducing Florida's pollution. The state's ample sunshine could cleanly and efficiently power both homes and cars.

HYDROGEN

Hydrogen is an almost completely clean-burning gas, it can be easily produced by running an electric current through water, and when burned it releases none of the carbon that leads to global warming. Furthermore, it can be derived from renewable resources. Of course, it does take energy to separate hydrogen from those elements it is naturally com-

Living
As If
There
Were A
Tomorrow

bined with, but by using solar power it can be accomplished without creating pollutants.

Its widespread use depends on the development of cost-effective methods to produce, store, and use the fuel. Hydrogen already powers the space shuttle; perhaps it will someday power our cars.

COMPRESSED NATURAL GAS

The cleanest-burning of the fossil fuels, natural gas releases significantly less harmful exhaust emissions than gas or diesel when used to power vehicles. This could make compressed natural gas (CNG) an important part of the solution to Florida's growing air pollution problems. Because CNG is cheaper than gasoline, and CNG-powered vehicles get about the same number of miles per gallon as similar gasoline powered vehicles, this alternative fuel offers significant cost savings as well.

WHAT TO DO

- Support alternative energy programs.
- Urge your representatives to support legislation promoting renewable energy resources.
- Buy solar-powered products, such as battery chargers, calculators, flashlights and radios.
- Install a solar water heating system.
- If you're building a new home or renovating an existing building, incorporate passive solar design features.

RESOURCES

FLORIDA SOLAR ENERGY CENTER
   300 State Rd. 401
   Cape Canaveral, FL 32920
   (407) 783-0300
   Call or write for a list of consumer information circulars.

REAL GOODS ALTERNATIVE ENERGY SOURCE BOOK
   Ten Speed Press
   P. O. Box 7123
   Berkeley, CA 94707
   800-841-2665

   This combination catalog, reference book and instruction manual makes sense of solar, wind and hydroelectric technology and offers all the equipment you'll need to make yourself energy self-sufficient. The cost is $14 plus $1.50 shipping.

REAL GOODS TRADING COMPANY
966 Mazzoni St.
Ukiah, CA 95482
FAX: 707-468-0301

Sells alternative energy products.

GOVERNOR'S ENERGY INFORMATION OFFICE
214 S. Bronough St.
Tallahassee, FL 32399-0001
(904)488-6764

Call or write for information about state energy programs.

INSTITUTE OF FOOD AND AGRICULTURAL
SCIENCES ENERGY PROGRAMS
Rolfs Hall, Room 220
University of Florida
Gainesville, FL 32611
(904)392-5240 or 392-5684

Call or write for information about programs.

FLORIDA CONSERVATION FOUNDATION, INC.
Environmental Information Center
1191 Orange Ave.
Winter Park, FL 32789
(305)644-5377 or 644-0641

INTERNATIONAL ASSOCIATION FOR HYDROGEN ENERGY
P.O. Box 248266
Coral Gables, FL 33124
(305)284-4667

FLORIDA ENVIRONMENTS, October 1991, Pg.17.

Living
As If
There
Were A
Tomorrow

DON'T BE A
SLAVE TO
AIR-CONDITIONING

Residential energy use accounts for close to 25 percent of total energy use in Florida, and almost half of this is for cooling our homes. Good air circulation and shade can counter the effects of heat and high humidity, and lessen our dependence on air-conditioning. Here are some ways to save energy and money.

- Take advantage of natural ventilation. Open your windows during the three or four months of the year when outside temperatures and humidity are at comfortable levels.

- Use ceiling fans. The cooling effect they produce allows the thermostat to be raised as much as 4 degrees. Moving the thermostat from 78 to 81 degrees can reduce cooling costs by one-fifth.

- Use a whole-house fan to lower attic temperatures and create a wind-chill effect. This works well during spring and fall seasons, as well as summer morning and evening hours. Whole house fans should not be used, however, when humidity is high. If they are, wood and upholstery will absorb moisture, making your air conditioner work harder when you do turn it on.

- Consider installing window awnings. During the period of the day when the sun shines directly on a South-facing window, a canvas awning will reduce the heat entering your home by 55 to 65 percent. On a West-facing window the reduction is between 72 and 77 percent. Metal awnings are even more efficient with heat gain reductions between 70 and 75 percent on South-facing windows, and between 75 and 80 percent on West-facing windows.

- Place sun control films on the outside of your windows. The more reflective the film is, the more effective it will be at keeping heat out.

- Plant shade trees so that they shade windows and glass doors throughout the day, as well as walls facing East and West. South-facing walls will benefit from shade as well, though they have less exposure to the sun. This is because during Florida's summers the sun rises in the northeast and sets in the northwest. By August, however, the sun drops low enough in the sky to significantly increase the afternoon heat load on a south facing wall.

TAKE A RATIONAL APPROACH TO RADON

You can't see it or smell it but, when inhaled, this naturally occurring radioactive gas can cause lung cancer. Florida is one of 12 states with significant health-threatening levels of radon. Outdoors it poses no problem because it dissipates. It can enter a home through leaks in the foundation or water pipes, and become a major health hazard in homes with little natural air exchange.

The Environmental Protection Agency rates radon as the greatest public health risk in water. Though drinking water with radon in it is generally safe for consumption, the radon becomes dangerous when it's released into the air. This can take place in the shower or from boiling water.

There are ways to deal with excess radon. If it's entering through cracks in the foundation, it can be sealed out. Filtration or aeration can remove radon if it's entering through the water.

WHAT TO DO

- Test the air in your home for radon. Radon air testers are available at hardware stores for $15 to $50.

- If the test results are greater than four and your water comes from a well, check your water.

- Treat your water if it tests high, then check your air again.

- Seal cracks in the foundation with patching concrete.

RESOURCES

RADON—A HOMEOWNER'S GUIDE TO DETECTION AND CONTROL
by Dr. Bernard S. Cohen and the editors of Consumer Reports Books

RADON REDUCTION METHODS and A CITIZEN'S
GUIDE TO RADON

Both available from:
Environmental Protection Agency
Public Information Center, PM211b
Washington, D.C. 204601

Living
As If
There
Were A
Tomorrow

VOLUNTEER OPPORTUNITIES

"More powerful than the will to win is the courage to begin."

—*Anonymous*

In order to make some environmental actions effective, people must reach out to others. You may need to join a group, begin a group, or persuade government leaders to make significant changes. In this chapter, you will find tips for becoming a mover and shaker for Florida's environment.

SUPPORT AN EXISTING PROJECT

Consider joining one of Florida's existing environmental groups or the Florida chapter of a national group. By joining one or more of these groups, you can provide financial support for various environmental issues. Many environmental memberships include a regular newsletter to keep you posted on the group's activities and emerging issues. Most groups will gladly provide a summary of how their funds are spent if you are interested.

If you want to work on specific local issues, look for a local group. Check with your local chamber of commerce office or newspaper for a directory of local clubs and organizations. Most newspapers publish an ongoing list of local meetings. Don't be bashful. Groups love to have new members who are enthusiastic and ready to work.

RESOURCES

See appendices for Florida Environmental Organizations.

ORGANIZE A GROUP

"Never doubt that a small group of thoughtful,
committed citizens can change the world; indeed,
it's the only thing that ever has."

—Margaret Mead, *Anthropologist*

Before starting from scratch with a new group, research existing groups and state organizations that might like to have a local chapter in your area. Is there an organization already addressing your concern? Are they effective? If there is not a group effectively addressing your pet issue, consider forming a group.

HOW TO FORM A GROUP

PLAN

Sometimes a crucial environmental issue will emerge almost overnight and citizens will have to work quickly to organize with little time for planning. However, a little planning in the beginning can make for a much more successful group.

- Locate a core group of members. network with friends, family, and community members to locate a small group of people who are interested in the same environmental issue.

- Define the organizational goals. develop one to three clearly defined goals that can be discussed and finalized at the first official meeting.

- Develop a list of potential group names. Based on the group's goals, brainstorm possible names with members of the core group. It is better to make the name positive rather than negative or hostile-sounding. Take the list to the first meeting for more brainstorming and a final vote.

- Establish first meeting date and agenda. Where and when you have your meetings is very important. Try to select a place that is easy to find and offers a comfortable meeting space. The first meeting is crucial. Prepare handouts for the group with proposed goals, suggested group names, brief background on the issue(s) addressed and key telephone contacts.

ORGANIZE

Volunteer Opportunities

- Incorporation and Tax-Exempt Status. If a group is going to be long-lived, raise a lot of money or enter into law-suits, it should consider becoming incorporated. Sometimes groups can recruit an attorney who will offer to file incorporation papers and provide other *pro bono* services.

GROW

- Expect group conflict. Conflict within groups is a natural first stage. Be prepared for differences of opinion and varying personalities.

- Keeping members involved. Offer specific and easy ways that people can help. Distribute a member form to collect vital information on each member and what resources and services they might offer the group. Use that form to establish phone committees, letter writers, lobbyists, researchers, etc.

- Reward and recognize achievement. Volunteers need to feel that they are appreciated and are making a differences. Set achievable goals. Celebrate small advances.

- Communicate regularly to members. Two of the most important tools for group communication are the telephone committee and the monthly newsletter.

- Develop a strong identity in the community and with the media. Research issues thoroughly. Give the other side a chance to correct their actions before calling in the press. Keep an open line of communication with both the media and opposing groups. Become a viable source of information to the media.

RESOURCES

FLORIDA ENVIRONMENTAL NETWORKS AND COALITIONS: CITIZEN ACTIVIST HANDBOOK and FLORIDA ENVIRONMENTAL NETWORKS AND COALITIONS: ORGANIZING GUIDE.
Florida Defenders of the Environment
2606 NW Sixth St.
Gainesville, FL 32609
(904) 372-6965

Two terrific guides for new environmental groups.

CITIZENS HANDBOOK TO THE LOCAL COMPREHENSIVE PLANNING ACT, Casey and David Gluckman,
Florida Audubon Society
460 Hwy. 436, Suite 200
(407) 260-8300

ECONET
Institute for Global Communications
18 DeBloom St.
San Francisco, CA 94107
(415) 442-0220

A computer based information network for individuals and groups interested in protecting the environment.

ENVIRONMENTAL SUCCESS INDEX
Renew America
1400 16th St. NW, Suite 710
Washington, D.C. 20036
(202) 232-2252

Highlights successful environmental programs from throughout the United States. Cost: $25.

Volunteer Opportunities

LOBBYING

*"Verbal persuasion is much more
cost-effective than legal confrontation."*

—Casey and David Gluckman
*Citizen's Handbook to the Local
Government Comprehensive Planning Act*

Lobbying is not just for big businesses. Lobbying is a tool that anyone can use effectively to inform and persuade elected officials about an issue.

Lobbyists use three main communication methods to get their points across: face-to-face meetings, telephone calls and letters. Here a few tips on effective lobbying.

PERSONAL MEETINGS

One-to-one communication is the most powerful lobbying tool. Here are a few tips for making the most out of a personal contact with an elected official.

- Set up an appointment in advance whenever possible. Let the official know the subject of your meeting. He or she can have staff research the issue.

- Make a good first impression by dressing and acting like a professional.

- Carefully prepare a brief, factual presentation including both sides of the issue. Do not overstate the facts or become too emotional or angry.

- Act as a resource or problem solver for the official. Officials need citizen input. Offer solutions. Offer information.

- Listening is one of the most powerful skills in lobbying. Find out what the official knows about the issue. Listen for their key objections or concerns. Repeat them for clarification. ("Do I understand that you are concerned about this bill because.................?") Find out what information they need and try to get it for them. Offer alternatives, amendments, or compromise solutions.

- Follow-up in writing or in person with specific answers to their concerns. Have others write to support your side (See letters below).

PUBLIC MEETINGS

Public meetings offer a chance to get views before the media and other interested parties. However, to be an effective lobbyist, citizens must communicate with elected officials before such meetings.

Most decisions are not made at public meetings or hearings. Informed elected officials gather research and make decisions in advance of public meetings. No official wants to be put publicly on the spot. If the official seems agreeable to your position, use the meeting to praise them for their foresight and understanding. Do not use a public meeting to criticize or attack a public official. Keep presentations factual and to the point without attacking individuals.

- When making presentations for a group, encourage as many group members as possible to be present. As you introduce yourself and your organization, ask your members to stand to show their support.

- Try to do something during the presentation that is memorable or newsworthy. One citizen activist complaining about stormwater pollution around a pristine Florida river brought a jar of nasty water she collected going into the river.

- Enlarged photos, charts, newsclippings and other visual aids help draw attention to your cause.

- Keep the points simple and summarize them in the end. Prepare a brief handout that people at the meeting can take with them.

TELEPHONE CALLS

Officials do keep score of calls on heated issues. Telephone calls do count. The best time to contact a state agency is before the bill is into final draft. The best time to contact an elected official is just before a vote on the issue. Most of Florida's major environmental groups have newsletters that alert citizens about upcoming legislation.

Some groups send out "legislative alerts" on important issues. To keep informed, become an active member of one of more of the groups interested in the same environmental issues you are. If the elected official is not available and time is running short before an important vote, leave a brief message stating who you are and that you are urging him or her to support bill #____ or such-and-such issue.

Volunteer
Opportunities

- For groups, consider using a legislative telephone network. Have each person contact three others to ask them to write or call about an important legislative issue.

LETTERS

HERE ARE A FEW TIPS FOR EFFECTIVE LETTER WRITING.

- Petitions and form letters are often ignored. So write personal letters.

- Keep letters brief, to the point and personal. One page is long enough. Handwritten is fine.

- Address one issue or bill per letter, when possible. Identify the bill number and/or name of the bill and your specific recommendation.

- Explain why you are concerned about the issue. If it will negatively affect a lot of people, explain why.

- Ask for a response on where he or she stands. Include a self-addressed envelope.

- Remember to thank officials that help. Thank them personally and publicly.

RESOURCES

WHO'S WHAT: YOUR POCKET PICTORIAL GUIDE TO FLORIDA GOVERNMENT
Florida Trend State Government Bookshelf
800-288-7363

A pocket reference with up-to-date names, addresses and phone numbers of Florida's government officials. Cost: $6.

FLORIDA LEGISLATIVE INFORMATION GUIDE
Florida Trend Government Bookshelf
800-288-7363

A spiral-bound source book to help you become an effective lobbyist. Helps you track legislation. Cost: $12.50

THE FLORIDA DIRECTORY
Florida Communications Network, Inc.
P.O. Box 2099
Gainesville, FL 32602

Includes all the key people involved in decision-making at the state and local level, including 67 counties and the major branches of government plus a bonus section on Florida's Congressional Delegation. Cost: $99

FEDERAL LEGISLATIVE PHONE NUMBERS

National Wildlife Federation Legislative Hotline
(202) 797-6655

US Capitol Switchboard (will connect you with any Senate, House, or Congressional committee office)
(202) 224-3121

House and Senate Bill Status
(202) 225-1772

THE NATIONAL WILDLIFE FEDERATION'S ACTIVIST KIT
National Wildlife Federation Dept. 318
1400 16th St. N.W.
Washington, D.C. 20036-2266
800-432-6564

Includes directory of congressional members and their committees, tips for effective lobbying, and other useful information. Request item #79957. Cost: $4.95.

THE NATIONAL WILDLIFE FEDERATION'S 1991 CONSERVATION DIRECTORY
Same Address As Above

Lists more than 1900 environmental contacts including congressional committees and their members, related U.S. government agencies, citizen's groups (from international to regional), and important conservation institutions throughout the world. Cost: $18 plus $3.95 shipping.

RENEW AMERICA
1400 16 St. NW, Suite 710
Washington, D.C. 20036

Watches the legislature and produces a "Renew America Scorecard" revealing their voting patterns on environmental issues.

Volunteer
Opportunities

INVOLVING THE MEDIA

Support from the media can make a big difference to the success of any environmental project or issue. Grassroots communication can be very effective, but the media allows you to reach a lot more people a lot faster. If the environmental issue centers around a corporate polluter, the fairest approach is to contact that entity first and suggest that they clean up their act. If they fail to take action, then you should inform the press. The media seems to be very interested in environmental news. Most large papers and tv stations have resources or staff dedicated to environmental reporting. They are looking for good, factual stories that have a broad interest. Here are a few tips for utilizing the media in our quest to "protect paradise."

- Select a designated spokesperson who is knowledgeable and makes a good appearance on television.
- Develop an effective media list including all print and electronic media in your community or region. Don't overlook weekly newspapers, shoppers, and magazines.
- Prepare a good fact sheet with vital statistics and details supporting your stand, concern or project.
- Deliver information in-person when possible.
- Offer to appear on radio and tv talk shows.
- Build a rapport with local reporters and news people. Become a source of information for them. Keep them posted on any new occurrences. Invite them to key meetings. Send them your newsletter. Ask them to cover clean-up days or other interesting hands-on activities. Search out interesting interviews for them with key environmentalists.
- Check into free studio and air time on local cable stations.

RESOURCES

1992 FLORIDA NEWS MEDIA/VALUABLE SOURCES GUIDE
Florida Trend Executive Bookshelf
800-288-7363
Names, addresses, telephone and fax numbers for Florida's dailies, weeklies, radio and tv stations, and magazines. Cost: $34.95

COUNTY ENVIRONMENTAL CONTACTS

ALACHUA COUNTY
Office of Environmental Protection
1 S.W. Second Place
Gainesville, FL 32601
(904) 336-2442

BAKER COUNTY
Planning and Zoning Department
55 N. Third St.
MacClenny, FL 32063
(904) 259-3613

BAY COUNTY
Environmental Services Department
225 McKenzie Ave.
Panama City, FL 32401
(904) 784-4070

BRADFORD COUNTY
Environmental Services Department
P. O. Box 1148
Starke, FL 32091
(904) 964-4454

BREVARD COUNTY
Environmental Services Department
Suite 347, Building C
Brevard County Government Center
2725 St. John's St.
Melbourne, FL 32940
(407) 633-2003

BROWARD COUNTY
Environmental Services Department
2401 North Powerline Rd.
Pompano Beach, FL 33069
(305) 960-3190

CALHOUN COUNTY
Planning and Zoning Department
425 E. Central Ave.
Blountstown, FL 32424
(904) 574-5553

CHARLOTTE COUNTY
Planning and Zoning Department
18500 Murdock Circle
Port Charlotte, FL 33948-1094
(813) 743-1222

CITRUS COUNTY
Environmental Health Office
1300 S. Lecanto Highway
Lecanto, FL 32661
(904) 746-5885

CLAY COUNTY
Planning and Zoning Department
P. O. Box 367
Green Cove Springs, FL 32043
(904) 284-6375

COLLIER COUNTY
Environmental Services Department

3301 Tamiami Trail, East
Naples, FL 33962
(813) 774-8428

COLUMBIA COUNTY
Environmental Services Department
249 E. Franklin St.
Lake City, FL 32055
(904) 755-4100

DADE COUNTY
Department of Environmental Resource
Management
111 N.W. First St., Suite 1310
Miami, FL 33128-1971
(305) 375-3376

DESOTO COUNTY
Planning and Zoning Department
104 N. Volusia Ave.
Arcadia, FL 33821
(813) 494-0042

DIXIE COUNTY
Environmental Health Department
P. O. Box 2099
Cross City, FL 32628
(904) 498-5732

DUVAL COUNTY
Bio-Environmental Services Department
421 W. Church St., Suite 412
Jacksonville, FL 32206
(904) 630-3638

ESCAMBIA COUNTY
Environmental Services Department
1190 W. Leonard St., Suite 2
Pensacola, FL 32501
(904) 444-8990

FLAGLER COUNTY
Planning and Zoning Department
P. O. Box 787
Bunnell, FL 32110
(904) 437-2218

FRANKLIN COUNTY
Environmental Services Department
P. O. Box 340
Apalachicola, FL 32320
(904) 653-9783

GADSDEN COUNTY
Environmental Services Department
P. O. Box 1799
Quincy, FL 32351
(904) 875-4448

GILCHRIST COUNTY
Planning and Zoning Department
P. O. Box 739
Trenton, FL 32693
(904) 463-6311

Environmental
Organizations
and Agencies

GLADES COUNTY
Planning and Zoning Department
P. O. Box 1018
Moore Haven, FL 33471
(813) 946-1525

GULF COUNTY
Planning and Zoning Department
1000 Fifth St.
Port St. Joe, FL 32456
(904) 229-6112

HAMILTON COUNTY
Board of Commissioners
P. O. Box 312
Jasper, FL 32052
(904) 792-1288

HARDEE COUNTY
Planning and Zoning Department
413 W. Orange St.
Wauchula, FL 33873
(813) 773-3236

HENDRY COUNTY
Planning and Zoning Department
P. O. Box 1760
LaBelle, FL 33935-1760
(813) 675-5220

HERNANDO COUNTY
Environmental Health Department
602 W. Broad St.
Brooksville, FL 34601
(904) 754-4072

HIGHLANDS COUNTY
Planning and Zoning Department
P. O. Box 1926
Sebring, FL 33871
(813) 382-5226

HILLSBOROUGH COUNTY
Environmental Protection Commission
P. O. Box 1110
Tampa, FL 33601
(813) 272-5960

HOLMES COUNTY
Environmental Services Department
201 N. Oklahoma St.
Bonifay, FL 32425
(904) 547-3691

INDIAN RIVER COUNTY
Planning and Zoning Department
1840 25 St.
Vero Beach, FL 32960
(407) 567-8000

JACKSON COUNTY
Environmental Services Department
P. O. Box 310
Marianna, FL 32446
(904) 526-2412

JEFFERSON COUNTY
Environmental Services Department
Jefferson County Courthouse, Room 10
Monticello, FL 32344
(904) 997-3573

LAFAYETTE COUNTY
Board of Commissioners
P. O. Box 88
Mayo, FL 32066
(904) 294-1600

LAKE COUNTY
Environmental Services Department
315 W. Main St.
Tavares, FL 32778
(904) 343-9739

LEE COUNTY
Environmental Services Department
1831 Hendry St.
Fort Myers, FL 33901
(813) 335-2477

LEON COUNTY
Environmental and Growth Management
Department
3401 W. Tharpe St.
Tallahassee, FL 32303
(904) 488-9300

LEVY COUNTY
Planning and Zoning Department
P. O. Box 672
Bronson, FL 32621
(904) 486-4311

LIBERTY COUNTY
Planning and Zoning Department
c/o Heniger & Ray, Inc.
630 E. Highway 44
Crystal River, FL 23629
(904) 795-6551

MADISON COUNTY
Board of Commissioners
P. O. Box 237
Madison, FL 32240
(904) 973-3179

MANATEE COUNTY
Environmental Services Department
P. O. Box 1000
Bradenton, FL 34206
(813) 745-3717

MARION COUNTY
Environmental Services Department
P. O. Box 2408
Ocala, FL 23678
(904) 629-0137

MARTIN COUNTY
Growth Management Department
2401 S.E. Monterey Rd.
Stuart, FL 34996
(407) 288-5495

MONROE COUNTY
Environmental Services Department
5100 Junior College Rd.
Public Services Building, Wing III
Key West, FL 33040
(305) 292-4406

NASSAU COUNTY
Planning and Zoning Department
2290 S. Eighth St.
Fernandina Beach, FL 32034
(904) 261-3706

OKALOOSA COUNTY
Environmental Services Department
84 Ready Ave.
Fort Walton Beach, FL 32548
(904) 651-3710

OKEECHOBEE COUNTY
Planning and Zoning Department
303 N.W. Second St.
Okeechobee, FL 34972
(813) 763-5548

ORANGE COUNTY
Environmental Protection Department
2002 E. Michigan Ave.
Orlando, FL 32806
(407) 836-7400

OSCEOLA COUNTY
Environmental Services Department
17 S. Vernon Ave., Room 155
Kissimmee, FL 34741
(407) 847-1200

PALM BEACH COUNTY
Environmental Resource Management
Department
3111 S. Dixie Highway, Suite 146
West Palm Beach, FL 33405
(407) 355-4011

PASCO COUNTY
Environmental/Fiscal Services Department
7536 State St.
New Port Richie, FL 34654
(813) 847-8145

PINELLAS COUNTY
Environmental Management Department
315 Court St.
Clearwater, FL 34616
(813) 462-4761

POLK COUNTY
Environmental Services Department
P. O. Box 39
Bartow, FL 33830
(813) 533-1205

PUTNAM COUNTY
Planning and Zoning Department
P. O. Box 758
Palatka, FL 32178
(904) 329-0310

SANTA ROSA COUNTY
Board of Commissioners
801 Caroline St., S.E., Suite J
Milton, FL 32570-4978
(904) 623-0135

SARASOTA COUNTY
Environmental Services Department
P. O. Box 8
Sarasota, FL 34230
(813) 364-4400

SEMINOLE COUNTY
Environmental Services Department
3000 A Southgate Dr.
Sanford, FL 32773
(407) 323-9615

ST. JOHNS COUNTY
Growth Management Department
P. O. Drawer 349
St. Augustine, FL 32085
(904) 823-2474

ST. LUCIE COUNTY
Growth Management Department
2300 Virginia Ave.
Fort Pierce, FL 34982
(407) 468-1590

SUMTER COUNTY
Public Services Department
209 N. FL St.
Bushnell, FL 33513
(904) 793-0270

SUWANNEE COUNTY
Board of Commissioners
200 S. Ohio Ave.
Live Oak, FL 32060
(904) 362-4002

TAYLOR COUNTY
Planning and Zoning Department
P. O. Box 620
Perry, FL 32347
(904) 584-6413

UNION COUNTY
Board of Commissioners
Union County Courthouse, Room 103
Lake Butler, FL 32054
(904) 496-1026

VOLUSIA COUNTY
Environmental Management Department
123 W. Indiana Ave.
DeLand, FL 32720-4612
(904) 736-5927

WAKULLA COUNTY
Comprehensive Planning Department
P. O. Box 1263
Crawfordville, FL 32327
(904) 599-8600

WALTON COUNTY
Comprehensive Planning Department
P. O. Box 689
DeFuniak Springs, FL 32433
(904) 892-7417

WASHINGTON COUNTY
Board of Commissioners
201 W. Chipley Ave., Suite 2
Chipley, FL 32428-1821
(904) 638-6200

**Environmental
Organizations
and Agencies**

REGIONAL ENVIRONMENTAL AGENCIES

REGIONAL PLANNING COUNCILS

Apalachee Regional Planning Council
314 E. Central Ave., Room 119
Blountstown, FL 32424
(904) 674-4571

Central Florida Regional Planning Council
P. O. Drawer 2089
Bartow, FL 33830
(813) 533-7130

East Central Florida Regional
Planning Council
1011 Wymore Rd., Suite 105
Winter Park, FL 32789
(407) 623-1075

North Central Florida Regional Planning
Council
235 S. Main St., Suite 205
Gainesville, FL 32601-6294
(904) 336-2200

Northeast Florida Regional Planning council
8649 Baypine Rd., Suite 110
Jacksonville, FL 32216
(904) 730-6260

South Florida Regional Planning Council
3440 Hollywood Blvd., Suite 140
Hollywood, FL 33021
(305) 961-2999

Southwest Florida Regional Planning Council
P. O. Box 3455
North Fort Myers, FL 33918-3455
(813) 995-4282

Tampa Bay Regional Planning Council
9455 Koger Blvd., Suite 219
St. Petersburg, FL 34900
(813) 577-5151

Treasure Coast Regional Planning Council
P. O. Box 1529
Palm City, FL 33490
(407) 286-3313

West Florida Regional Planning Council
P. O. Box 486
Pensacola, FL 32593-0386
(904) 444-8910

Withlacoochee Regional Planning Council
1241 Southwest 10th St.
Ocala, FL 32674-2798
(904) 732-1315

Florida Regional Planning Councils Association
902 N. Gadsden St.
Tallahassee, FL 32303
(904) 224-1020

WATER MANAGEMENT DISTRICTS

Northwest Florida Water Management District
Rt. 1, Box 3100
Havana, FL 32333
(904) 539-5999

St. Johns River Water Management District
P. O. Box 1429
Palatka, FL 32078-1429
(904) 329-4500

South Florida Water Management District
P. O. Drawer 24680
West Palm Beach, FL 33416
(407) 686-8800

Southwest Florida Water Management District
2379 Broad St.
Brooksville, FL 34609-6899
(904) 796-7211

Suwannee River Water Management District
Rt. 3, Box 64
Live Oak, FL 32060
(904) 362-1001

STATE ENVIRONMENTAL AGENCIES

DEPARTMENT OF AGRICULTURE
The Capitol
Tallahassee, FL 32399-0810
(904) 488-3022

Division of Forestry
3125 Conner Blvd.
Tallahassee, FL 32399-1650
(904) 488-8466

DEPARTMENT OF COMMUNITY AFFAIRS
2740 Centerview Dr
Tallahassee, FL 32399-2100
(904) 488-8466

DEPARTMENT OF EDUCATION

Office of Environmental Education
Room 224-C, Florida Education Center
Tallahassee, FL 32399-0040
(904) 487-7900

**DEPARTMENT OF
ENVIRONMENTAL REGULATION**
2600 Blair Stone Rd.
Tallahassee, FL 32399-2400
(904) 488-4805

**DEPARTMENT OF HEALTH AND REHABILITA-
TIVE SERVICES**
1317 Winewood Blvd.
Tallahassee, FL 32399-0700
(904) 488-7721

Office of Environmental Health
(904) 488-6811

DEPARTMENT OF NATURAL RESOURCES
3900 Commonwealth Blvd.
Tallahassee, FL 32399-3000
(904) 488-1554

DEPARTMENT OF STATE
500 South Bronough St.
R. A. Gray Building
Tallahassee, FL 32399-0250
(904) 488-3680

Bureau of Archeological Research
(904) 487-2299

Bureau of Archives and Records Management
(904) 487-2073

DEPARTMENT OF TRANSPORTATION

Clean Florida Commission
Burns Building, Mail Drop 2
Tallahassee, FL 32399-0450
(904) 488-3111

EXECUTIVE OFFICE OF THE GOVERNOR

Environmental Policy/Community and
Economic Development Unit
Office of Planning and Budgeting
Executive Office of the Governor
The Capitol
Tallahassee, FL 32399-0001
(904) 488-5551

Office of Cabinet Affairs
Executive Office of the Governor
The Capitol
Tallahassee, FL 32399-0001
(904) 488-5152

Florida/Washington, D.C. Office
444 N. Capitol St., Suite 287
Washington, D.C. 20001
(202) 624-5885

**FLORIDA ADVISORY COUNCIL ON
ENVIRONMENTAL EDUCATION**
Room 237, Holland Building
Tallahassee, FL 32399-1400
(904) 487-0123

FLORIDA HOUSE OF REPRESENTATIVES

Natural Resources Committee
House Office Building, Room 202
Tallahassee, FL 32399-1300
(904) 488-1564

FLORIDA SENATE

Natural Resources and
Conservation Committee
Senate Office Building, Room 414
Tallahassee, FL 32399-1100
(904) 487-5372

GAME AND FRESH WATER FISH COMMISSION
620 South Meridian St.
Tallahassee, FL 32399-1600
(904) 488-6661

MARINE FISHERIES COMMISSION
2540 Executive Center Circle, West
Suite 106
Tallahassee, FL 32301
(904) 487-0554

Environmental
Organizations
and Agencies

FLORIDA ENVIRONMENTAL GROUPS AND ORGANIZATIONS

AMERICAN LITTORAL SOCIETY
Coral Reef Conservation Center
75 Virginia Beach Dr
Key Biscayne, FL 33149
(305) 361-4495

AMERICAN RIVERS CONSERVATION
COUNCIL, FLORIDA OFFICE
927 Delores Dr
Tallahassee, FL 32301
(904) 942-0414

APALACHEE LAND CONSERVANCY
2424 West Lakeshore Dr
Tallahassee, FL 32303
(904) 385-7997

CENTER FOR MARINE CONSERVATION
One Beach Dr , Southeast
Suite 304
St. Petersburg, FL 33701
(813) 895-2188

CITIZENS FOR A BETTER SOUTH FLORIDA
2025 Southwest 32 Ave.
Miami, FL 33145-2211
(305) 444-9555

CLEAN WATER ACTION
1320 South Dixie Highway
Coral Gables, FL 33146
(305) 661-6992

CLEAN WATER FUND
1320 South Dixie Highway
Coral Gables, FL 33146
(305) 661-6992

COASTAL ADVISORY COMMITTEE
1068 Jackpine St.
West Palm Beach, FL 33413
(407) 793-3954

COASTAL PLAINS INSTITUTE
1313 North Duval St.
Tallahassee, FL 32301
(904) 681-6208

THE CONSERVANCY, INC.
1450 Merrihue Dr
Naples, FL 33942
(813) 263-0067

DOLPHIN PROJECT
P. O. Box 224
Coconut Grove, FL 33233
(305) 443-9012

MARJORY STONEMAN DOUGLAS
BISCAYNE NATURE CENTER
34 LaGorce Circle
Miami, FL 33141
(305) 861-2899

BIG BEND EARTH FIRST
P. O. Box 20582
Tallahassee, FL 32316
(904) 224-6782

ENVIRONMENTAL EDUCATION
FOUNDATION OF FLORIDA, INC.
Suite 1106, The Capitol
Tallahassee, FL 32399-0001
(904) 488-5551

ENVIRONMENTAL SOLUTIONS
INTERNATIONAL
P. O. Box 869
Islamorada, FL 33036
(305) 664-9796

ENVIRONMENTAL STUDIES CENTER
P. O. Box 636
Gonzalez, FL 32560
(904) 968-0135

EVERGLADES COORDINATING COUNCIL
610 Northwest 93 Ave.
Pembroke Pines, FL 33024
(305) 653-1416

FLORIDA AUDUBON SOCIETY
460 Hwy 436, Suite 200
Casselberry, FL 32707
(407) 260-8300

FLORIDA CENTER FOR SOLID AND HAZ-
ARDOUS WASTE MANAGEMENT
University of Florida
3900 Southwest 63 Blvd.
Gainesville, FL 32608-3848
(904) 392-6264

FLORIDA CONSERVATION ASSOCIATION
904 East Park Ave.
Tallahassee, FL 32301
(904) 681-0438

FLORIDA CONSERVATION FOUNDATION
1251-B Miller Ave.
Winter Park, FL 32789
(407) 644-5377

FLORIDA DEFENDERS OF THE ENVIRON-
MENT
2606 Northwest 6th St.
Gainesville, FL 32609
(904) 372-6965
FAX (904) 377-0869

FLORIDA FEDERATION OF GARDEN CLUBS
P. O. Box 1604
Winter Park, FL 32790-1604
(407) 647-7016

FLORIDA FORESTRY ASSOCIATION
P. O. Box 1696
Tallahassee, FL 32302
(904) 222-5646

FLORIDA GEOLOGICAL SURVEY
Department of Natural Resources
903 West Tennessee St.
Tallahassee, FL 32304-7795
(904) 488-5380

FLORIDA KEYS LAND AND SEA TRUST
P. O. Box 536
Marathon, FL 33050
(305) 743-3900

FLORIDA LEAGUE OF CONSERVATION
VOTERS
P. O. Box 857
Floral City, FL 32636
(904) 726-1566

FLORIDA NATIVE PLANT SOCIETY
P. O. Box 680008
Orlando, FL 32868
(407) 299-1472

FLORIDA NATURAL AREAS INVENTORY
1018 Thomasville Rd., Suite 200-C
Tallahassee, FL 32303
(904) 224-8207

FLORIDA PUBLIC INTEREST RESEARCH
GROUP
308 E. Park Ave., Suite 213
Tallahassee, FL 32301
(904) 224-5304

FLORIDA RESOURCES NETWORK
290 N.E. 131st St.
N. Miami, FL 33161
(305) 895-0787 (305) 759-4478

FLORIDA SEA GRANT COLLEGE
University of Florida
Building 803, IFAS 0341
Gainesville, FL 32611-0341
(904) 392-5870

FLORIDA SHORE AND BEACH PRESERVATION
ASSOCIATION, INC.
864 East Park Ave.
Tallahassee, FL 32301
(904) 222-7677

FLORIDA SOLAR ENERGY CENTER
300 State Road 401
Cape Canaveral, FL 32920
(407) 783-0300

FLORIDA TRAIL ASSOCIATION
P.O. Box 13708
Gainesville, FL 32604
(904) 378-8823

FLORIDA WILDLIFE FEDERATION
P.O. Box 6870
Tallahassee, FL 32314
(904) 656-7113

FLORIDA WILDLIFE REHABILITATOR'S
ASSOCIATION
c/o Miami Museum of Science
3280 South Miami Ave.
Miami, FL 33129-9989
(305) 854-4247 Extension 244

FLORIDA WILDLIFE SANCTUARY
2600 Otter Creek Ln.
Melbourne, FL 32935
(305) 254-8843

FRIENDS OF THE EVERGLADES
101 Westward Dr., Suite 2
Miami Springs, FL 33166
(305) 888-1230

FRIENDS OF THE OLETA RIVER
1011 Northeast 141 St.
North Miami, FL 33161
(305) 895-0787

FRIENDS OF THE WEKIVA
164 Wekiva Park Dr
Sanford, FL 32771
(407) 321-3848

GOPHER TORTOISE COUNCIL
P. O. Box 61301
St. Petersburg, FL 33784-1301
(813) 327-3779

GULF COAST ZOOLOGICAL SOCIETY
Palm Aire Plaza
5899 Whitfield Ave., Number 205
Bradenton, FL 34243
(813) 355-0000

HOBE SOUND NATURE CENTER
P. O. Box 214
Hobe Sound, FL 33475
(407) 546-2067

IZAAK WALTON LEAGUE OF AMERICA, INC.
Florida Division
31 Garden Cove Dr.
Key Largo, FL 33037
(305) 451-0993

JOINT CENTER FOR ENVIRONMENTAL AND
URBAN PROBLEMS
1515 West Commercial Blvd.
Fort Lauderdale, FL 33309
(305) 776-1240

KEEP FLORIDA BEAUTIFUL
402 West College Ave.
Tallahassee, FL 32301
(904) 561-0700

LEGAL ENVIRONMENTAL ASSISTANCE FOUN-
DATION (LEAF)
1115 N. Gadsden St.
Tallahassee, FL 32303-6327
(904) 681-2591

LEAGUE OF ENVIRONMENTAL
EDUCATORS IN FLORIDA
c/o Suwannee River Water Management
District
Rt. 3, Box 64
Live Oak, FL 32060
(904) 362-1001

MANASOTA-88
5314 Bay State Rd.
Palmetto, FL 34221
(813) 722-7413

MARINE RESOURCES COUNCIL
150 W. University Blvd.
Melbourne, FL 32901
(407) 768-8000

MARINE SPILL RESPONSE CORPORATION
Southeast Region
905 S. American Way
Miami, FL 33132
(305) 375-8410

**Environmental
Organizations
and Agencies**

MUSEUM OF SCIENCE AND INDUSTRY
4801 E. Fowler Ave.
Tampa, FL 33617
(813) 985-1914

THE NATIONAL AUDUBON SOCIETY
Southeast Regional Office
928 North Munroe St.
Tallahassee, FL 32303
(904) 222-2473

NATIONAL PARKS AND CONSERVATION
ASSOCIATION
Florida Office
126 Mill Branch Road
Tallahassee, FL 32312
(904) 893-4959

THE NATURE CONSERVANCY
Florida Regional Office
2699 Lee Road, Suite 500
Winter Park, FL 32789
(407) 628-5887

THE NATURE CONSERVANCY
Tallahassee Office
515-A North Adams St.
Tallahassee, FL 32301
(904) 222-1099

OCTAGON WILDLIFE SANCTUARY
41660 Horseshoe Rd.
Punta Gorda, FL 33955
(813) 543-1130

1000 FRIENDS OF FLORIDA
524 East College Ave.
Tallahassee, FL 32301
(904) 222-6277

ORGANIZED FISHERMEN OF FLORIDA
P. O. Box 740
Melbourne, FL 32902
(407) 725-5212

PEOPLE ORGANIZED TO
PREVENT POLLUTION
P. O. Box 3290
Spring Hill, FL 34606
(813) 856-1196

PINE JOG ENVIRONMENTAL
EDUCATION CENTER
6301 Summit Blvd.
West Palm Beach, FL 33415
(407) 686-6600

RAILS -TO-TRAILS CONSERVANCY
Florida Chapter
2545 Blair Stone Pines Dr
Tallahassee, FL 32301
(904) 942-2379

REEF RELIEF
P. O. Box 470
Key West, FL 33041
(305) 294-3100

ST. FRANCIS WILDLIFE REHABILITATION
AND EDUCATION CENTER
1521 Fuller Rd.
Tallahassee, FL 32303
(904) 222-8436

SANIBEL-CAPTIVA CONSERVATION
FOUNDATION, INC.

P. O. Box 839
3333 Sanibel-Captiva Rd.
Sanibel, FL 33957
(813) 472-1932

SAVE THE MANATEE CLUB
500 North Maitland Ave.
Maitland, FL 32751
(407) 539-0990

SAVE WHAT'S LEFT
Coral Springs High School
7201 W. Sample Rd.
Coral Springs, FL 33065
(305) 344-3412

SIERRA CLUB
Florida Chapter
1201 North Federal Highway
North Palm Beach, FL 33408
(407) 775-3846

SUWANNEE RIVER COALITION
4025 S.W. 18th St.
Gainesville, FL 32608

TALL TIMBERS RESEARCH STATION
Rt. 1, Box 678
Tallahassee, FL 32312
(904) 893-4153

TALLAHASSEE JUNIOR MUSEUM
3945 Museum Dr.
Tallahassee, FL 32310
(904) 575-8684

TROPICAL AUDUBON SOCIETY
5530 Sunset Dr.
South Miami, FL 33143
(305) 666-5111

TRUST FOR PUBLIC LAND
1310 Thomasville Rd.
Tallahassee, FL 32303-5608
(904) 222-9280

WEKIVA RESOURCES COUNCIL
University of Central Florida
Orlando, FL 32816-0150
(407) 275-2942

THE WILDERNESS SOCIETY
Regional Headquarters
4203 Ponce deLeon Blvd.
Coral Gables, FL 33146
(305) 448-3636

WILDLIFE ALERT
Florida Game and Fresh Water Fish Commission
620 South Meridian St.
Tallahassee, FL 32399-1600
(800) 342-1676 (Panama City)
(800) 342-8105 (Lake City)
(800) 282-8002 (Lakeland)
(800) 342-9620 (Ocala)
(800) 432-2046 (West Palm Beach)

WILDLIFE SOCIETY
Florida Chapter
56CSS/DEN
Avon Park Air Force Range
Avon Park, FL 33825
(813) 452-4119

INDEX

Environmental Organizations and Agencies